Gene Therapy in Reconstructive and Regenerative Surgery

Giorgio Giatsidis
Editor

Gene Therapy in Reconstructive and Regenerative Surgery

Springer

Editor
Giorgio Giatsidis
Division of Plastic Surgery, Department of Surgery
Brigham and Women's Hospital, Harvard Medical School
Boston, MA
USA

ISBN 978-3-319-78956-9 ISBN 978-3-319-78957-6 (eBook)
https://doi.org/10.1007/978-3-319-78957-6

Library of Congress Control Number: 2018942886

Printed on acid-free paper

This Springer imprint is published by the registered company Springer International Publishing AG part of Springer Nature.
The registered company address is: Gewerbestrasse 11, 6330 Cham, Switzerland

To my mom and my dad, selfless and dedicated parents and doctors, for constantly inspiring me with their unique example to do my best and positively impact other lives.

Contents

Principles of Gene Therapy in Reconstructive and Regenerative Surgery

1

Giorgio Giatsidis

1.1 Advantages of Gene Therapy Approaches in Reconstructive and Regenerative Surgery

Gene therapy is defined as the introduction of a genetic material into cells to alter cellular functions or structures with the goal of improving a clinical outcome: this approach initially focused on the correction of inherited systemic diseases [2]. Despite encouraging results, the treatment of systemic genetic disorders with gene therapy still represents a challenging task, due to the difficulty in providing a long-term, sustained, and widespread expression of edited genes in target tissues. These limitations only minimally impact the therapeutic potential of gene therapy in reconstructive and regenerative surgeries, as in most of these cases a short-term, temporary, and localized (topical) gene expression/induction is sufficient to achieve successful clinical outcomes. Traumatic or iatrogenic/surgically induced tissue injuries are an example of this scenario. Gene-edited cells instructed to temporarily express proangiogenic factors might enhance and accelerate healing of traumatic cutaneous injuries, or the transfection of antifibrotic genes could help reduce the impact of scarring during wound healing in both physiologic and pathologic (e.g., burns, keloids) conditions [4–8]. Similarly, temporary gene therapies could be effectively adopted in pediatric patients for the treatment/correction of congenital malformations (e.g., craniofacial, upper limb) during development [9]. Conversely, long-term gene induction could be preferable for the treatment of prolonged/chronic disorders. Examples of these could include chronic wounds caused

G. Giatsidis
Division of Plastic Surgery, Department of Surgery, Brigham and Women's Hospital, Harvard Medical School, Boston, MA, USA
e-mail: ggiatsidis@bwh.harvard.edu

G. Giatsidis (ed.), *Gene Therapy in Reconstructive and Regenerative Surgery*,
https://doi.org/10.1007/978-3-319-78957-6_1

by diabetes or sustained pressure in hospitalized patients and paraplegics [10, 11]. In reconstructive and regenerative surgery, there is usually a need to apply treatments only locally in a well-defined area of the body This favors gene therapy approaches by enabling the use of several methods for the direct application of therapeutic agents to the affected tissue. In addition, gene therapy techniques have less side effects than other advanced biologic therapies, such as those based on the administration of growth factors or stem cells. Because of their critical role in reparative and regenerative processes, growth factors have been extensively used in the treatment of wounds and tissue injuries [12]. Differently from the direct application of growth factors to tissues, gene transfer usually allows for a longer modulation of the same factors in target tissues, a higher bioavailability, a slower inactivation, and lower costs for prolonged applications [2]. In addition, in contrast to other (localized or systemic) treatments, gene expression can be modulated over time by controlling the magnitude of its activity, temporarily halting or reversing it, or suppressing it. Growth factors are not the only target for gene therapy in tissue repair, and some promising new approaches have recently been reported using transcription factors, transduction molecules (e.g., SMAD family), RNA interference (iRNA), and host defense peptides (HDP) as therapeutic agents delivered by gene therapies [10, 11, 13, 14].

1.2 Standard Methods for Gene Delivery to Tissues

Genes can be delivered to a target tissue using either ex vivo or in vivo methods [2, 3, 15]. Both strategies have advantages and drawbacks; ongoing research efforts are continuously improving efficacy, effectiveness, and safety of these methods.

Ex vivo methods involve the isolation of cells from the patient and their in vitro transfection before in vivo grafting. This approach is commonly requires more time- and resource-consuming but it also often results in a more efficient outcome.

In vivo methods involve the direct delivery of a gene into a target cell or tissue. Technically easier and requiring less resources (e.g., a GMP cell facility), this approach is usually less specific and efficient.

In both cases, the same transfection systems can be used to introduce therapeutic genes in a target cell (in vitro or in vivo): these systems can be divided into viral and non-viral methods.

1.2.1 Viral Transfection Techniques

Viruses are natural vehicles for gene delivery and have been the first to be adopted in early gene therapy research [8, 16–20]. Viral vectors are constructed by deleting genes essential for virus replication and replacing them with sequences encoding a gene of interest. Retroviruses, adenoviruses, adeno-associated viruses, and herpes simplex virus are the most tested viral gene delivery systems, and they have been widely used to repair tissues. Distinct advantages and disadvantages control the applicability of each viral system to gene therapies. These include the size of the gene to be inserted, the characteristic of the target cells (dividing or non-dividing), the need for long-term expression, the immunogenicity, and the ability for genomic integration. Despite few disadvantages, viral vectors remain a preferred method for gene transfer in preclinical research, given their capacity to achieve prolonged expression and elevated transfection efficiency (Table 1.1).

1.2.2 Non-Viral Transfection Techniques

Non-viral gene transfection is associated with a lower inflammatory response, a reduced risk of infections, and more simple protocols of manipulation. However, this approach usually yields less specific and more variable gene expression in target tissues. Non-viral gene transfer techniques includes naked DNA transfection and liposomal transfection [21–26].

Since direct injection of DNA into tissues has shown to be largerly ineffective, naked DNA transfer is usually achieved through the adoption of facilitating techniques, such as micro-seeding, "gene guns," ultrasounds, microbubbles, or electroporation. Gene electro-transfer designates the combination of gene transfer and electroporation.

Liposomal transfection, instead, uses cationic liposomes: these are synthetically prepared vesicles-with a positively charged surface-that contain the cDNA of a gene

Table 1.1 Key properties of the most adopted viral vectors in reconstructive gene therapies

Virus	Advantages	Disadvantages
Retrovirus	Transfection of different cell type; high efficiency of ex vivo transfection; induction of long-term gene expression	Transfection only of dividing cells; inefficient transfection in vivo; risks of insertional mutagenesis
Adenovirus	Transfection of virtually all cell types (dividing and nondividing cells); good efficiency of in vivo transfection; no integration into host genome	Risks of inducing immune response; lack of permanent expression; potential wild-type breakthrough; small size of DNA insert (8 kb)
Adeno-associated virus	Transfection of virtually all cell types (dividing and nondividing cells); integration in specific site (chromosome 19); induction of long-term gene expression	Risks of insertional mutagenesis; small size of DNA insert (4.7 Kb); risks of inducing immune response or inflammatory reactions; difficulties producing high titers
Herpes simplex virus-1	Transfection of different call types; neurotropism; large size of DNA insert (30 kb); large DNA insert size (30 kb); induction of long-tern gene expression	Difficult to manipulate because of complex life cycle; potential wild-type breakthrough

Reproduced from: The role of gene therapy in regenerative surgery: updated insights. Giatsidis G, Dalla Venezia E, Bassetto F. Plast Reconstr Surg. 2013 Jun;131 (6):1425–35 of interest. Liposomes can be applied by either topical administration or by direct injection, but only the latter approach has shown effective rates of cell transfection and gene expression. Several groups have demonstrated that liposomes containing DNA sequences for growth factors enhance, accelerate, and improve dermal and epidermal regeneration.

1.3 Combination of Tissue Engineering Approaches and Gene Transfer Techniques

Substantial interest and research have been focused in recent years on the integration of gene therapies with classic tissue-engineered approaches, such as those involving the development of regenerative scaffolds, the use of stem cells, or combinations of both these two. It has been shown that integrating strategies can be synergistically beneficial, result in significantly improved outcomes and decrease complications or limits of each single approach [18, 27–31].

Bioengineered scaffolds can incorporate a plasmid DNA and modulate its release either directly (polymeric release) or through a substrate-mediated delivery [2]. Polymeric release methods incorporate viral vectors in a scaffold and control their release by varying the degradation kinetics of the scaffold. Instead, substrate-mediated delivery immobilizes vectors on the surfaces of scaffolds for direct exposure to the surrounding cellular environment, hence spatially controlling the transduction of a therapeutic agent. Scaffold-incorporated and scaffold-released gene transfer has shown to prolong transgenic expression, protect DNA from degradation and immune response, and extend opportunities for cell transfection. The release of DNA by biomaterials can also take advantage of both viral and non-viral methods.

Stem cells, due to their pluripotency and their growth potential, represent an ideal method for transferring genes to tissues [18, 32]. In addition, their intrinsic regenerative abilities can be further increased by the transfection of therapeutic genes, creating "boosted" regenerative stem cells. Recent discoveries on the low immunogenicity of specific sub-populations of stem cells (e.g., mesenchymal) suggest that also allogenic or xenogenic cells could be used as vectors for therapeutic genes.

1.4 Target Tissues for Gene Therapy Approaches in Reconstructive and Regenerative Surgery

Every tissue that is repaired by reconstructive surgeries, can be modified or regenerated using gene transfer approaches.

The skin has been the first, most popular, and easiest target of gene therapies [2, 3, 6, 8, 11, 33]. Besides the fact that most medical conditions in reconstructive and regenerative surgery involve the skin, making it a priority to develop improved treatments, what makes the skin a favorable tissue for genetic manipulation is its accessibility, ease of monitoring for side effects and adverse reactions, and its ease of transfection. Fibroblasts and keratinocytes, the predominant cells of the skin, are easily harvested and cultured, allowing in vitro amplification and genetic manipulation. The high turnover and self-renewing capacity of the epidermis also make it an ideal target for gene delivery. Long-term gene expression in keratinocytes has proven to be challenging until very recently, and the lower turnover in fibroblasts makes the process of transfection more difficult. Yet, in most clinical cases, transient gene expression may be sufficient or desirable in treating acute cutaneous conditions such as wound repair, infections, or localized inflammatory processes. Wounds (acute/chronic), burn injuries, and pathologic scarring have been areas of major interest for the application of gene therapies to the skin [2, 3, 6, 8, 11, 33].

Gene therapy of bone injuries has also been a popular target of research [2, 3, 34, 35]. The possibility to deliver osteo-inductive factors (e.g., bone morphogenetic proteins) to stimulate bone repair and remodeling has been investigated in several preclinical studies. Given the widespread development of tissue-engineered scaffolds (e.g., demineralized bone matrix, collagen, PLGA, or hydroxyapatite) and the availability of stem cell therapies to promote bone repair, the combination of gene therapies with both these other therapeutic approaches has also been widely explored.

Articular cartilage is a remarkably durable tissue, but once injured, it has very limited capacities for self-repair due to the poor blood supply. Repair responses, in most cases, result in a fibrocartilagineous tissue with inferior mechanical and biochemical characteristics. This challenge has stimulated research to develop therapeutic approaches aimed to achieve a more functional/physiologic repair of injured cartilage by transplanting genetically modified chondrocytes in damaged tissues [2, 3, 36, 37].

Tendon healing is a process that lasts several months, culminating in a disorganized tissue architecture with reduced elasticity and mobility that increases the risk of injury recurrence. As current reparative/regenerative therapies have offered only suboptimal outcomes, increasing research has focused on the use of recombinant cytokines and growth factors, delivered to injured tendons using gene therapies, often combined with (stem/) cells or scaffolds [2, 3, 19, 38–41].

Gene transfer of neurotrophic factors, mostly in combination with cell therapies and scaffolds (nerve conduits), has also been investigated as a method to promote nerve repair and stimulate the regenerative capacity of peripheral nerves [2, 3, 42–45].

Muscle and soft tissue flaps represent the workhorse of modern reconstructive surgery. Despite tremendous advances in the last decades, flap surgeries are not free from complications, mostly related to the impaired survival of transferred tissues after the surgically induced hypoxia. Several studies have investigated the possibility to modify flaps using gene therapies to make them more resilient to ischemic insults and improve surgical outcomes (e.g., transfection of DNA encoding for pro-angiogenic growth factors, such as VEGF) [2, 23, 46–50]. Another area of research has focused on the use of flaps as scaffolds for the delivery of gene therapies to tissues. Examples of this research include the development of flaps that are resistant to radiation therapy, that produce anti-bacterial factors, or that can secrete antitumorigenic factors. [51–54].

Conclusions

Despite being over two decades old, the field of gene therapy seems to have seen an abrupt acceleration toward clinical application only very recently. Translation of experimental protocols into clinical trials is still a long, expensive, and challenging process. So far, in the field of reconstructive and regenerative surgery, clinical evidence of the effectiveness of gene therapies has been provided only by few, yet promising, clinical trials and sparse case reports. Even so, there is a growing enthusiasm in the field, fueled by recent technological advances and medical breakthroughs that have been achieved in the application of gene therapies to other fields of medicine [55]. Translation of these therapies to patient care will need to address the potential risks intrinsically associated with the treatment (e.g., use of viral vectors, suboptimal control of levels of protein expression and targeted transfection) and with gene-editing. In 2012, gene therapy for reconstructive and regenerative surgery was defined as "*an adolescent looking forward to growing up*": [2] 6 years later, the adolescent has grown into a young adult, almost ready to transform standards of care in this field, for the benefit of a large number of patients.

References

1. Hirsch T, Rothoeft T, Teig N, et al. Regeneration of the entire human epidermis using transgenic stem cells. Nature. 2017;551(7680):327–32. https://doi.org/10.1038/nature24487.
2. Giatsidis G, Dalla Venezia E, Bassetto F. The role of gene therapy in regenerative surgery: updated insights. Plast Reconstr Surg. 2013;131(6):1425–35. https://doi.org/10.1097/PRS.0b013e31828bd153.
3. Tepper OM, Mehrara BJ. Gene therapy in plastic surgery. Plast Reconstr Surg. 2002;109(2):716–34. http://www.ncbi.nlm.nih.gov/pubmed/11818860. Accessed January 12, 2018.
4. Baisch A, Riedel F. Hyperplastic scars and keloids: part II: surgical and non-surgical treatment modalities. HNO. 2006;54(12):981–92-4. https://doi.org/10.1007/s00106-006-1463-y.
5. Xu B, Liu ZZ, Zhu GY, et al. Efficacy of recombinant adenovirus-mediated double suicide gene therapy in human keloid fibroblasts. Clin Exp Dermatol. 2008;33(3):322–8. https://doi.org/10.1111/j.1365-2230.2007.02615.x.
6. Mofazzal Jahromi MA, Sahandi Zangabad P, Moosavi Basri SM, et al. Nanomedicine and advanced technologies for burns: preventing infection and facilitating wound healing. Adv Drug Deliv Rev. 2018;123:33–64. https://doi.org/10.1016/j.addr.2017.08.001.
7. Mendoza-Garcia J, Sebastian A, Alonso-Rasgado T, Bayat A. Optimization of an ex vivo wound healing model in the adult human skin: functional evaluation using photodynamic therapy. Wound Repair Regen. 2015;23(5):685–702. https://doi.org/10.1111/wrr.12325.
8. Balaji S, Lesaint M, Bhattacharya SS, et al. Adenoviral-mediated gene transfer of insulin-like growth factor 1 enhances wound healing and induces angiogenesis. J Surg Res. 2014;190(1):367–77. https://doi.org/10.1016/j.jss.2014.02.051.
9. Tollefson TT, Senders CW, Sykes JM. Changing perspectives in cleft lip and palate: from acrylic to allele. Arch Facial Plast Surg. 2008;10(6):395–400. https://doi.org/10.1001/archfaci.10.6.395.
10. Icli B, Nabzdyk CS, Lujan-Hernandez J, et al. Regulation of impaired angiogenesis in diabetic dermal wound healing by microRNA-26a. J Mol Cell Cardiol. 2016;91:151–9. https://doi.org/10.1016/j.yjmcc.2016.01.007.
11. Steinstraesser L, Lam MC, Jacobsen F, et al. Skin electroporation of a plasmid encoding hCAP-18/LL-37 host defense peptide promotes wound healing. Mol Ther. 2014;22(4):734–42. https://doi.org/10.1038/mt.2013.258.

12. Golas AR, Hernandez KA, Spector JA. Tissue engineering for plastic surgeons: a primer. Aesthet Plast Surg. 2014;38(1):207–21. https://doi.org/10.1007/s00266-013-0255-5.
13. Whittam AJ, Maan ZN, Duscher D, et al. Challenges and opportunities in drug delivery for wound healing. Adv Wound Care. 2016;5(2):79–88. https://doi.org/10.1089/wound.2014.0600.
14. Wackerhage H, Ratkevicius A. Signal transduction pathways that regulate muscle growth. Essays Biochem. 2008;44(1):99–108. https://doi.org/10.1042/bse0440099.
15. Ghali S, Dempsey MP, Jones DM, Grogan RH, Butler PE, Gurtner GC. Plastic surgical delivery systems for targeted gene therapy. Ann Plast Surg. 2008;60(3):323–32. https://doi.org/10.1097/SAP.0b013e31806917b0.
16. Liao J, Wei Q, Fan J, et al. Characterization of retroviral infectivity and superinfection resistance during retrovirus-mediated transduction of mammalian cells. Gene Ther. 2017;24(6):333–41. https://doi.org/10.1038/gt.2017.24.
17. Yi CG, Xia W, Zhang LX, et al. VEGF gene therapy for the survival of transplanted fat tissue in nude mice. J Plast Reconstr Aesthet Surg. 2007;60(3):272–8. https://doi.org/10.1016/j.bjps.2006.01.052.
18. Sharma P, Wimalawansa SM, Gould GC, Johnson RM, Excoffon KJDA. Adeno-associated virus 5 transduces adipose-derived stem cells with greater efficacy than other adeno-associated viral serotypes. Hum Gene Ther Methods. 2016;27(6):219–27. https://doi.org/10.1089/hgtb.2016.123.
19. Tang JB, Chen CH, Zhou YL, McKeever C, Liu PY. Regulatory effects of introduction of an exogenous FGF2 gene on other growth factor genes in a healing tendon. Wound Repair Regen. 2014;22(1):111–8. https://doi.org/10.1111/wrr.12129.
20. Jackson JD, McMorris AM, Roth JC, et al. Assessment of oncolytic HSV efficacy following increased entry-receptor expression in malignant peripheral nerve sheath tumor cell lines. Gene Ther. 2014;21(11):984–90. https://doi.org/10.1038/gt.2014.72.
21. Qiu L, Zhang L, Wang L, et al. Ultrasound-targeted microbubble destruction enhances naked plasmid DNA transfection in rabbit Achilles tendons in vivo. Gene Ther. 2012;19(7):703–10. https://doi.org/10.1038/gt.2011.165.
22. Delalande A, Gosselin M-P, Suwalski A, et al. Enhanced Achilles tendon healing by fibromodulin gene transfer. Nanomedicine. 2015;11(7):1735–44. https://doi.org/10.1016/j.nano.2015.05.004.
23. Liu PY, Tong W, Liu K, et al. Liposome-mediated transfer of vascular endothelial growth factor cDNA augments survival of random-pattern skin flaps in the rat. Wound Repair Regen. 2004;12(1):80–5. https://doi.org/10.1111/j.1067-1927.2004.012114.x.
24. Hirsch T, Spielmann M, Velander P, et al. Insulin-like growth factor-1 gene therapy and cell transplantation in diabetic wounds. J Gene Med. 2008;10(11):1247–52. https://doi.org/10.1002/jgm.1251.
25. Seyed Jafari SM, Shafighi M, Beltraminelli H, Geiser T, Hunger RE, Gazdhar A. Improvement of flap necrosis in a rat random skin flap model by in vivo electroporation-mediated HGF gene transfer. Plast Reconstr Surg. 2017;139(5):1116e–27e. https://doi.org/10.1097/PRS.0000000000003259.
26. Rezende FC, Gomes HC, Lisboa B, Lucca AF, Han SW, Ferreira LM. Electroporation of vascular endothelial growth factor gene in a unipedicle transverse rectus abdominis myocutaneous flap reduces necrosis. Ann Plast Surg. 2010;64(2):242–6. https://doi.org/10.1097/SAP.0b013e318196cbe0.
27. Chávez MN, Schenck TL, Hopfner U, et al. Towards autotrophic tissue engineering: photosynthetic gene therapy for regeneration. Biomaterials. 2016;75:25–36. https://doi.org/10.1016/j.biomaterials.2015.10.014.
28. Perisic T, Zhang Z, Foehr P, et al. Biodegradable poly (lactic acid-co-glycolic acid) scaffolds as carriers for genetically-modified fibroblasts. PLoS One. 2017;12(4):e0174860. https://doi.org/10.1371/journal.pone.0174860.
29. Alluri R, Jakus A, Bougioukli S, et al. 3D printed hyperelastic "bone" scaffolds and regional gene therapy: a novel approach to bone healing. J Biomed Mater Res A. 2017;106(4):1104–10. https://doi.org/10.1002/jbm.a.36310.

30. Lu P, Zhang G-R, Cai Y-Z, et al. Lentiviral-encoded shRNA silencing of proteoglycan decorin enhances tendon repair and regeneration within a rat model. Cell Transplant. 2013;22(9):1507–17. https://doi.org/10.3727/096368912X661292.
31. Lu P, Zhang GR, Song XH, Zou XH, Wang LL, Ouyang HW. Col V siRNA engineered tenocytes for tendon tissue engineering. PLoS One. 2011;6(6):e21154. https://doi.org/10.1371/journal.pone.0021154.
32. Nixon AJ, Watts AE, Schnabel LV. Cell- and gene-based approaches to tendon regeneration. J Shoulder Elb Surg. 2012;21(2):278–94. https://doi.org/10.1016/j.jse.2011.11.015.
33. Mo Y, Guo R, Zhang Y, Xue W, Cheng B, Zhang Y. Controlled dual delivery of angiogenin and curcumin by electrospun nanofibers for skin regeneration. Tissue Eng Part A. 2017;23(13–14):597–608. https://doi.org/10.1089/ten.tea.2016.0268.
34. Bougioukli S, Sugiyama O, Pannell W, et al. Gene therapy for bone repair using human cells: superior osteogenic potential of BMP-2 transduced mesenchymal stem cells derived from adipose tissue compared to bone marrow. Hum Gene Ther. 2017. https://doi.org/10.1089/hum.2017.097.
35. Wu G, Zhou B, Hu C, Li S. Gene expression of osteogenic factors following gene therapy in mandibular lengthening. J Craniofac Surg. 2015;26(2):378–81. https://doi.org/10.1097/SCS.0000000000001085.
36. Zhang Q, Gong W, Ning B, Nie L, Wooley PH, Yang S-Y. Local gene transfer of OPG prevents joint damage and disease progression in collagen-induced arthritis. Sci World J. 2013;2013:1–8. https://doi.org/10.1155/2013/718061.
37. Tang Y, Wang B. Gene- and stem cell-based therapeutics for cartilage regeneration and repair. Stem Cell Res Ther. 2015;6(1):78. https://doi.org/10.1186/s13287-015-0058-5.
38. Wu YF, Mao WF, Zhou YL, Wang XT, Liu PY, Tang JB. Adeno-associated virus-2-mediated TGF-β1 microRNA transfection inhibits adhesion formation after digital flexor tendon injury. Gene Ther. 2016;23(2):167–75. https://doi.org/10.1038/gt.2015.97.
39. Tang JB, Zhou YL, Wu YF, Liu PY, Wang XT. Gene therapy strategies to improve strength and quality of flexor tendon healing. Expert Opin Biol Ther. 2016;16(3):291–301. https://doi.org/10.1517/14712598.2016.1134479.
40. Tang JB, Wu YF, Cao Y, et al. Basic FGF or VEGF gene therapy corrects insufficiency in the intrinsic healing capacity of tendons. Sci Rep. 2016;6(1):20643. https://doi.org/10.1038/srep20643.
41. Majewski M, Betz O, Ochsner PE, Liu F, Porter RM, Evans CH. Ex vivo adenoviral transfer of bone morphogenetic protein 12 (BMP-12) cDNA improves Achilles tendon healing in a rat model. Gene Ther. 2008;15(16):1139–46. https://doi.org/10.1038/gt.2008.48.
42. Hoyng SA, De Winter F, Gnavi S, et al. Gene delivery to rat and human Schwann cells and nerve segments: a comparison of AAV 1-9 and lentiviral vectors. Gene Ther. 2015;22(10):767–80. https://doi.org/10.1038/gt.2015.47.
43. Jiang W, Tang L, Zeng J, Chen B. Adeno-associated virus mediated SOD gene therapy protects the retinal ganglion cells from chronic intraocular pressure elevation induced injury via attenuating oxidative stress and improving mitochondrial dysfunction in a rat model. Am J Transl Res. 2016;8(2):799–810. http://www.ncbi.nlm.nih.gov/pubmed/27158370. Accessed January 12, 2018.
44. Zor F, Deveci M, Kilic A, et al. Effect of VEGF gene therapy and hyaluronic acid film sheath on peripheral nerve regeneration. Microsurgery. 2014;34(3):209–16. https://doi.org/10.1002/micr.22196.
45. Busuttil F, Rahim AA, Phillips JB. Combining gene and stem cell therapy for peripheral nerve tissue engineering. Stem Cells Dev. 2017;26(4):231–8. https://doi.org/10.1089/scd.2016.0188.
46. O'Toole G, MacKenzie D, Lindeman R, et al. Vascular endothelial growth factor gene therapy in ischaemic rat skin flaps. Br J Plast Surg. 2002;55(1):55–8. https://doi.org/10.1054/bjps.2001.3741.
47. Moimas S, Manasseri B, Cuccia G, et al. AAV vector encoding human VEGF165-transduced pectineus muscular flaps increase the formation of new tissue through induction of angiogenesis in an in vivo chamber for tissue engineering: a technique to enhance tissue and vessels

in microsurgically engineered tissue. J Tissue Eng. 2015;6:2041731415611717. https://doi.org/10.1177/2041731415611717.

48. Huang N, Khan A, Ashrafpour H, et al. Efficacy and mechanism of adenovirus-mediated VEGF-165 gene therapy for augmentation of skin flap viability. Am J Physiol Heart Circ Physiol. 2006;291(1):H127–37. https://doi.org/10.1152/ajpheart.01253.2005.
49. Jung H, Gurunluoglu R, Scharpf J, Siemionow M. Adenovirus-mediated angiopoietin-1 gene therapy enhances skin flap survival. Microsurgery. 2003;23(4):374–80. https://doi.org/10.1002/micr.10140.
50. Michaels J, Dobryansky M, Galiano RD, et al. Ex vivo transduction of microvascular free flaps for localized peptide delivery. Ann Plast Surg. 2004;52(6):581–4. http://www.ncbi.nlm.nih.gov/pubmed/15166989. Accessed January 12, 2018.
51. Dempsey MP, Hamou C, Michaels J, et al. Using genetically modified microvascular free flaps to deliver local cancer immunotherapy with minimal systemic toxicity. Plast Reconstr Surg. 2008;121(5):1541–53. https://doi.org/10.1097/PRS.0b013e31816ff6aa.
52. Agrawal VK, Copeland KM, Barbachano Y, et al. Microvascular free tissue transfer for gene delivery: in vivo evaluation of different routes of plasmid and adenoviral delivery. Gene Ther. 2009;16(1):78–92. https://doi.org/10.1038/gt.2008.140.
53. Michaels J, Levine JP, Hazen A, et al. Biologic brachytherapy: ex vivo transduction of microvascular beds for efficient, targeted gene therapy. Plast Reconstr Surg. 2006;118(1):54–65-8. https://doi.org/10.1097/01.prs.0000220466.27521.22.
54. Ghali S, Bhatt KA, Dempsey MP, et al. Treating chronic wound infections with genetically modified free flaps. Plast Reconstr Surg. 2009;123(4):1157–68. https://doi.org/10.1097/PRS.0b013e31819f25a4.
55. Pineda M, Moghadam F, Ebrahimkhani MR, Kiani S. Engineered CRISPR systems for next generation gene therapies. ACS Synth Biol. 2017;6(9):1614–26. https://doi.org/10.1021/acssynbio.7b00011.

Gene Therapy in Skin and Wound Healing

2

Kristo Nuutila, Mansher Singh, and Elof Eriksson

2.1 Introduction

Wound healing is a complex process governed by several different signaling pathways. In acute wounds tissue destruction triggers acute and chronic inflammatory cascades, epidermal repair, dermal repair, and a prolonged process of remodeling/maturation. Chronic wounds display signs of being trapped in a prolonged inflammatory phase [1–3]. Since Cohen's discovery of epidermal growth factor in 1962, thousands of in vitro and animal studies have described the signaling proteins and pathways linking them together in wound healing [4, 5]. A few clinical studies have also been done with only one growth factor, platelet-derived growth factor (PDGF), showing clinical effect [6, 7].

Gene transfer (often more pretentiously referred to as gene therapy) implies the insertion of DNA or RNA into the target cells. The goal is either to express a missing or under-expressed protein or to block a deleterious protein [8]. There are pathophysiologic difficulties. The wound healing processes have a large amount of overlapping redundancies where no single signaling protein regulates a process or

K. Nuutila (✉)
Division of Plastic Surgery, Department of Surgery, Brigham & Women's Hospital, Harvard Medical School, Boston, MA, USA

Faculty of Medicine, Department of Pharmacology, University of Helsinki, Helsinki, Finland

M. Singh
Department of General Surgery, Beth Israel Deaconess Medical Center and Harvard Medical School, Boston, MA, USA
e-mail: msingh6@bidmc.harvard.edu

E. Eriksson
Harvard Medical School, Boston, MA, USA
e-mail: elof@comcast.net

G. Giatsidis (ed.), *Gene Therapy in Reconstructive and Regenerative Surgery*,
https://doi.org/10.1007/978-3-319-78957-6_2

even part of a process [9]. It is akin to looking for a river to control its flow and instead finding a delta with many interconnected branches. To date a large number of signaling proteins have been shown to have a significant influence on the part of the wound healing process but not of sufficient magnitude to change the total outcome [10]. We are still looking for an extracellular or intracellular protein that is a critical regulator of the wound healing processes. That protein could then be expressed or blocked in a tightly regulated fashion with gene therapy.

Gene delivery offers many advantages over the therapeutic delivery of the recombinant protein. The DNA or RNA can be delivered to or close to the target cells. High expression of the protein can be achieved, and if the transfected cells are keratinocytes, these cells will undergo differentiation and eventually be eliminated from the skin. Inducible gene switches to turn on or off the transgene can also be employed [11, 12].

The vectors used for gene transfer can generally be divided into viral or non-viral. Modified retrovirus, adenovirus, adeno-associated virus, lentivirus, and herpes simplex virus are the most common. The virus has been made replication deficient, and the DNA or RNA and promoter sequences have been inserted into the virus. Non-viral vectors include naked DNA or RNA and liposomal constructs where the gene transfer process can be augmented by mechanical, electrical, or chemical means [13, 14].

2.2 Normal Wound Healing

The process of wound healing can be divided into three overlapping phases – inflammation, proliferation, and remodeling – that each wound needs to go through in order to heal normally. Inflammation phase occurs immediately and lasts for about 3 days. During this phase dead cells and pathogens are cleared out of the wound. The proliferation phase begins around 2 days after injury and continues until the wound is closed. In this phase growth of new tissue begins, and angiogenesis, collagen deposition, granulation tissue formation, wound contraction, and reepithelialization take place. The last phase, remodeling, occurs after the wound closure and can last more than a year. During this phase collagen realigns, all of the processes activated after the injury shut down, and cells that are no longer needed undergo apoptosis or exit from the wound [15].

2.3 Abnormal Wound Healing

The challenges in wound healing emerge when wounds fail to follow the path of normal healing process. Consequently, the wound healing is either downregulated (become chronic) or upregulated (scarring). The majority of wound healing problems are related to an excessive inflammatory process. Inflammation is an essential part of wound healing and crucial for the removal of infectious agents from the wound bed. However prolonged overexpression of pro-inflammatory cytokines (such as various interleukins, ILs, and tumor necrosis alpha (TNF-α)) in the wound inhibits the shift from inflammation phase to proliferation phase and delays the healing process; wounds underheal [16]. In addition, prolonged inflammation results in overexpression of matrix metalloproteinases (MMPs) in the wound. MMPs play an important role in wound repair by degrading and removing damaged extracellular matrix (ECM) molecules from the injured tissue [17]. Although being important in tissue regeneration, their excess proteolytic activity is associated with many disease processes such as chronic wounds because they destroy growth factors, cofactors, and cell surface receptors. If the inflammation phase continues for a longer period of time, the wounds underheal, become chronic, and fail to heal until the chronic inflammation is reduced. By comparing the wound exudate of chronic wounds to clean surgical wounds, it has been found that chronic wound exudate contains greater amounts of pro-inflammatory cytokines and proteases. In addition, growth factor levels are smaller in chronic wound exudate [18]. Thus, it has been suggested that lack of growth factors, imbalance between proteinases (MMPs) and their inhibitors (TIMPs), or presence of too many senescent cells in the injured area may cause inability of these wounds to heal [19].

When the acute inflammatory process is upregulated, it results in overhealing, and the wounds form excess fibrous connective tissue that results in scarring. This process is characterized by strong fibroblast proliferation that causes excessive growth of granulation tissue and wound contraction that ultimately lead to the formation of scar tissue [20]. The pathogenesis of fibrosis is a combination of several factors. An individual's genetic background, hormonal imbalance, and race greatly affect the prevalence of keloids and hypertrophic scars. However, in general it has been thought that prolonged inflammatory phase together with delayed onset of epithelialization causes scarring. In fact, increased amounts of pro-inflammatory mediators especially transforming growth factor beta (TGF-β) have been found in the wounds that heal by fibrosis [21].

2.4 Cutaneous Gene Transfer Techniques

There are two approaches to deliver genes into target tissue, either ex vivo or in vivo [22]. The skin is an excellent tissue for both techniques due to its easy accessibility that also allows the continuous monitoring of the effects of gene therapy on the skin. For ex vivo approaches, where the gene manipulation is done in vitro, both epidermal and dermal target cells are easy to isolate, culture, transfect, and transplant back to the recipient. The easy access to cells and efficient culturing protocols make the skin also an attractive tissue to test new gene therapy strategies in vitro. For in vivo approaches that are based on the direct introduction of genes to the target tissue, the skin is also suitable since gene transfer to the cells can be performed minimally invasively avoiding unnecessary systemic delivery, and the wound can be easily targeted repeatedly by various techniques, including injection, micro-seeding, and topical application [23]. Apart from the delivery approach, the success of the gene transfer to the target tissue ultimately depends on the DNA/RNA construct. The two major ways to transfer genetic material are viral vectors or non-viral methods (Fig. 2.1) [24].

2.4.1 Viral Gene Transfer

Viral gene transfer relies on the natural ability of viruses to penetrate into cells and express their genes in host cells. Various properties of the virus including viruses' immunogenicity, physical stability, size of the insert, and abilities to integrate into a host genome (transient/long-term expression) or infect target cells (dividing/nondividing cells) regulate their suitability as gene transfer vectors. The most commonly used viruses for gene delivery have been adenoviruses, adeno-associated viruses, retroviruses, and herpes simplex viruses [25].

Adenoviruses, medium-sized and naked, are composed of double-stranded DNA and most commonly used as vectors in gene therapy to express a foreign therapeutic gene [26]. Benefits of adenoviral vectors in wound healing gene therapy include high efficiency of transduction and the ability to infect both dividing and nondividing cells. In fact, adenoviruses have shown a 95% efficiency to infect skin cells in vitro [27]. The genetic material of adenoviruses does not integrate into the host genome resulting in transient expression of the introduced gene, which is of particular interest in gene therapy for wounds [28]. The biggest limitation on the use of

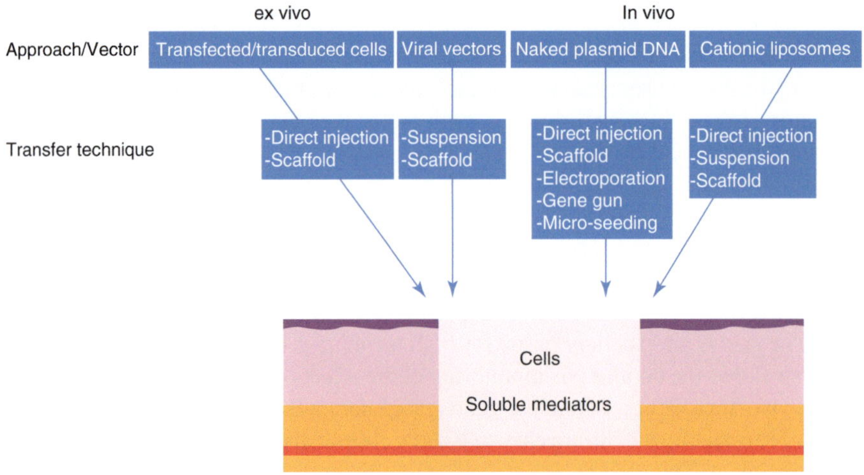

Fig. 2.1 There are two approaches to deliver genes into target tissue, either ex vivo or in vivo. In the ex vivo approach, the gene transfer is done in vitro to the target cells, and subsequently the genetically modified cells are being transplanted into the wound. The in vivo approaches are based on the direct introduction of genes to the target tissue. The two major ways to transfer genetic material are viral vectors or non-viral methods. Viral gene transfer relies on the natural ability of viruses to penetrate into cells and express their genes in host cells. Non-viral methods include various physical and chemical methods

adenoviral vectors has been the host cells' strong immune response to the adenoviral proteins that leads to local inflammation and destruction of the infected cells. However, many preclinical studies and clinical trials have indicated that adenovirus vectors are safe and have therapeutic activity [29].

Adeno-associated viruses (AAV) are small sized, naked, and composed of single-stranded DNA. They are not known to cause any disease, only a latent infection to host cells. These viruses have a stable viral capsid and thus low immunogenicity, which show almost no signs of immune response. Therefore, the main advantage of AAV vectors is safety [30]. Like adenoviral vectors, AAV vectors can infect both dividing and nondividing cells. In addition, these viruses integrate reliably into a specific site (chromosome 19) eliminating unforeseen mutagenesis of DNA. AAV vectors have shown transduction efficiency up to 70% for keratinocytes; in vitro and in vivo gene transfer into porcine skin has been attained with intradermal injection, which resulted in transgene expression in epidermal keratinocytes for more than 6 weeks. The disadvantages with AAVs are the small size of the vector, meaning that only short sequences of foreign DNA can be inserted into the host cell. In addition, these cells need helper viruses to proliferate, and usually after purification from the contaminating virus, the yield is low [28].

Retroviruses are medium-sized, enveloped single-stranded RNA viruses that replicate in a host cell through a process of reverse transcription. They have been largely used in various gene therapy clinical trials. Retroviruses have an ability to integrate into the host cell's genome providing long-term transgene expression. However, use of these viruses is limited by low transduction efficiency as well as their particle instability [31]. Lentiviruses, a subclass of retroviruses, are commonly used in gene therapy applications because unlike other retroviruses, they are able to infect both dividing and nondividing cells. Retroviral vectors have been used successfully ex vivo to transduce human growth hormones and growth factors to cultured keratinocytes [32].

Herpes simplex viruses (HSV) are large, enveloped, and composed of double-stranded DNA. HSV is very infectious and can efficiently infect numerous mammalian cells and tissue types, such as the skin, due to the wide expression pattern of the cellular receptors recognized by the virus [32]. HSV vectors have a very high transgenic capacity that allows them to insert long sequences of foreign DNA into a host cell. In addition, a long-term transgene expression can be achieved by using HSV vectors. The biggest concern in the use of HSV vectors is their high cytotoxicity. Therefore, the challenge has been to develop a vector that is safe [33, 34].

2.4.2 Non-viral Gene Transfer

Recently a trend for using non-viral vectors for gene transfer has increased. Concerns related to the safety of viral vectors have led to the development of several non-viral methods to transfer genes. The main advantage of non-viral gene transfer is its biosafety since these techniques do not cause any infection or additional inflammatory reactions [35]. Non-viral vectors also enable transfection of almost unlimited size of the transgene and repeated administration. Besides, the majority of the non-viral methods are simpler and cheaper than viral transfer because in most of the cases no ex vivo manipulation of the target cells is needed. A drawback has been the non-specificity of the transfection in terms of cells being targeted. The non-viral gene transfer systems can be divided into physical and chemical methods [36].

Physical methods for gene delivery utilize various mechanical forces to introduce transgene into cells. This approach is based on penetration through the cell membrane by causing transient pores through which DNA can transfect the cell [37]. Skin is a good organ for different physical gene transfer approaches due to its easy accessibility, and therefore numerous techniques, such as electroporation, gene gun, and micro-seeding, have been tried on wound healing [38].

In a method called *electroporation*, target cells have been suspended in a medium containing DNA plasmids. Subsequently, high-voltage electrical current is used to create temporary nanometric pores on the surface of the target cell allowing negatively charged DNA plasmids to move into cells [39]. This method has shown a lot of potential in wound healing and has been successfully used to accelerate chronic wound healing. The drawback of electroporation is low transfection efficiency [40].

The direct local injection of naked DNA has been applied on several tissues including the skin. In *gene gun technology* (also known as ballistic gene delivery or DNA particle bombardment), particles of heavy metals such as gold, silver, or tungsten (1–5 mm in diameter) are coated with plasmid DNA and then using a ballistic device driven into the tissue. The impulse of the particles enables a penetration about a few millimeters deep into the tissue, and DNA is released into skin cells as a result [41]. As a gene transfer technique, gene gun is relatively simple and safe since it doesn't elicit immunoresponse in the target cells. However, the downsides are low and transient transgene expression. In the context of wound healing, gene gun has been successfully used to transfer various growth factors to treat wounds [42].

Micro-seeding is another technique that transfers naked DNA directly into the target cells by multiple injections with a set of solid oscillating microneedles mounted on a modified tattooing machine. A plasmid DNA or vector solution is placed on the tips of the needles and introduced into to the skin or wound at a depth of 2–3 mm and at a rate of 7500 penetrations per 2.25 cm^2 surface area [42]. Using this method, the amount of plasmid DNA delivered can exceed 100 μg per target site, and increased expression of the transgene can be maintained 1–2 weeks. The disadvantage of micro-seeding is that it reaches only the superficial layers of the skin with minimal penetration into the deeper tissue [43].

Several charged *chemical compounds*, such as cationic lipids, different cationic polymers, and lipid polymers, can be used to transfect target cells. Cationic delivery techniques have many advantages compared to both other non-viral and viral gene transfer methods, such as low toxicity and antigenicity and long-term expression with less risk of insertional oncogenesis. However, low efficiency is also the weakness of cationic delivery systems. The most common *chemical gene transfer method* is to use *cationic liposomes* for gene transfer [44].

Cationic liposomes are positively charged vesicles made of lipids. The positive surface charge of the liposomes interacts with negatively charged molecules such as

DNA or RNA, forming a complex that enters the target cell through endocytosis. Due to the lack of immunogenicity, this technique also allows repeated transfections in vivo. Thus, cationic liposomes offer a simple and reproducible way to transfer large amounts of DNA, RNA, siRNA, or oligonucleotides into the target cells [45]. Liposomes can be administered to the wound topically or by direct injection. However, to intact skin direct injection is more efficient than topical administration due to the less permeable stratum corneum. Direct injection results in transient transfection of keratinocytes, fibroblasts, and inflammatory cells [46]. In wound healing cationic liposomes have been used in several preclinical studies to transfect various growth factors to the wound bed. The downside of cationic liposome gene transfer is low transfection efficiency in vivo in comparison to viral gene transfer techniques [47].

Scaffold-based gene delivery, in which transgenes are incorporated into scaffolds and encoded either by a viral or non-viral vector, is another attractive alternative for cutaneous gene transfer [48]. Introduction of a transgene can be done through a plasmid DNA incorporated into the scaffold or through genetically manipulated cells which have been seeded in vitro onto the scaffold. The scaffold is then transplanted at the site of injury where it also provides physical support for new tissue regeneration. The DNA is released from the scaffold either through hydrolysis or cellular migration. Numerous natural and synthetic materials such as hyaluronic acid, collagen, and chitosan have been used as matrices for *scaffold-based gene delivery* [49]. For instance, in a preclinical study, collagen matrix has been used to deliver an adenoviral vector encoding PDGF into the wound [50]. *Scaffold-based gene delivery* has several advantages such as increased residence time within the tissue, greater protection against degradation, and a reduced inflammatory response. The biggest challenges with this approach are related to inefficient delivery, lack of stable gene expression, inappropriate levels of gene expression, and immune clearance of either the vector or the cells expressing the foreign gene [45].

2.5 Regulation of Transgene Expression

The aim of gene therapy is to introduce genetic material into cells in the target tissue, which can lead to either addition or inhibition of a cellular function. In the cells, the transgene depending on the purpose can make cells produce beneficial protein to help fight a disease or heal a trauma, compensate for mutated genes that cause disease with a healthy copy of the gene, or knock out a mutated gene that has an undesired function (https://ghr.nlm.nih.gov/primer/therapy/genetherapy). In wound healing gene therapy is most commonly used to deliver genes into wounds to increase the production of growth factors, which accelerate healing. Unlike in some other gene therapy applications, in wound healing persistent expression of the transgene is not wanted. Once the wound has healed further, gene expression is not necessary, and therefore only transient transgene expression is preferred [51]. Non-viral gene transfer methods and adenoviral vectors are known for their transient but varying expressions in the skin and wounds, whereas most of the other viral vectors integrate stably into the host genome and thereby achieve long-term gene expression. Unnecessary and uncontrolled high transgene expression may lead to deleterious effects, and therefore, once the therapeutic gene has been successfully introduced into the target tissue, it is important to be able to control and regulate the expression of the foreign gene [12].

Several systems to control transgene expression (gene switches) have been developed, which are usually based on strategies where tissue-specific or inducible promoters (repressors) are inserted into recombinant vectors that can be regulated by a pharmacological agent. Systems based on various drugs such as tetracycline, rapamycin, progesterone, tamoxifen, streptogramin, macrolide, and ecdysone have been used [52]. As an example, Yao et al. demonstrated that tetracycline repressor-mediated (tetR) transcription repression system can function as a powerful gene switch for on-and-off regulation of therapeutic gene expression in ex vivo gene transfer protocols. Cells that express human endothelial growth factor (hEGF) under the control of the tetR-mediated transcription repression switch were transplanted into porcine full-thickness wounds. They showed that by topically applying tetracycline into wounds at various concentrations or at different time points posttransplantation, the levels and timing of hEGF expression in transplanted wounds could be reversibly regulated by the antibiotic. Tetracycline binds to the repressor, releasing the operator and thus initiating transcription [53].

In addition, transgene delivery using scaffolds offers a possibility for controlled release of the therapeutic gene. Biomaterial scaffolds can be loaded with vectors and engineered to release the transgene in a controlled manner by enhancing interactions between the scaffold and the vectors [54]. For instance, it has been shown that by incorporating affinity peptides into hydrogels, the vector release can be promoted or hindered [55].

2.6 Target Genes for Wound Healing

Skin and wound healing are great targets for gene therapy since they are easy to access, and the main cell types (keratinocytes and fibroblast) can be isolated, cultured, and genetically modified in the laboratory setting. Also importantly transient transgene expression is sufficient and even desirable for wound healing treatments [56].

During normal wound healing, the cells in the wound secrete various soluble mediators including growth factors, cytokines, chemokines, inflammatory factors, and extracellular matrix molecules. These molecules are all crucial for tissue healing by affecting several processes such as angiogenesis, granulation tissue formation, inflammation, and wound reepithelialization [57]. Disruption in the production of these factors may lead to prolonged inflammation, disturbed balance of proteases and their inhibitors, or inactivation of growth factors that further affect the wound healing [58]. Therefore, gene therapy offers a great tool to control the production of these factors by either increasing or inhibiting their expression during wound healing.

Majority of the gene therapy strategies to treat skin wounds have concentrated on increasing the production of growth factors. In addition, more recently small interfering RNA (siRNA) and microRNA (miRNA) have been delivered to wounds to silence and regulate gene expression. Various animal models with many different vectors have been used to study the potential of gene therapy on wound healing. In addition, some small clinical trials have been conducted in recent years [41, 45, 56, 59, 60]. Some preclinical and clinical gene therapy efforts and various target genes related to wound healing are described in the following chapters. Table 2.1 (preclinical) and Table 2.2 (clinical) summarize the studies depicted in the text.

Table 2.1 Table summarizes the preclinical studies (transgene, vector, animal model, transfer technique, and reference) depicted in the text

Transgene	Vector	Model	Transfer technique	References
VEGF	Viral: adeno-associated virus	Diabetic mouse/ full-thickness wound	Injection	[61]
VEGF	Viral: adeno-associated virus	Mouse/ partial-thickness burn	Injection	[62]
VEGF	Non-viral: minicircle plasmid DNA + PAM-RG4	Diabetic mouse/ full-thickness wound	Injection	[63]
FGF-1	Non-viral: cationic liposome	Diabetic mouse/ full-thickness wound	Topical	[64]
FGF-7(KGF)	Non-viral: cationic liposome	Rat/full-thickness wound	Injection	[65]
FGF-7(KGF)	Viral: adenovirus	Skin-humanized mouse model	Injection/fibrin matrix/ex vivo fibroblast	[66]
PDGF-AA	Viral: retrovirus	Athymic mice	Ex vivo keratinocytes	[67]
PDGF-B	Viral: adenovirus	Ischemic rabbit/ ear wound	Injection	[68]
EGF	Non-viral: plasmid DNA	Porcine/ partial-thickness wounds	Gene gun	[69]
EGF	Non-viral: plasmid DNA	Mouse/ full-thickness wound	Ex vivo fibroblasts	[70]
TGF-b	Non-viral: plasmid DNA	Rat/full-thickness wound	Gene gun	[71]
TGF-b	Non-viral: plasmid DNA	Diabetic mouse/ full-thickness wound	Topical (hydrogel)	[72]
HGF	Non-viral: plasmid DNA	Diabetic rat/ full-thickness wound	Non-needle jet injector	[73]

(continued)

Table 2.1 (continued)

Transgene	Vector	Model	Transfer technique	References
HGF	Viral: adenovirus	Rabbit/ear wound	Topical	[74]
IGF-1	Non-viral: cationic liposome	Diabetic porcine/ full-thickness	Ex vivo keratinocytes	[75]
IGF-1	Non-viral: cationic liposome	Rat/full-thickness wound	Injection	[76]
siRNA (target: Mapk-1)	Non-viral: cationic liposome	Mouse/ full-thickness wound	Topical (agarose matrix)	[77]
siRNA (target: PHD2)	Non-viral: endosomolytic nanoparticles	Mouse/ full-thickness wound	Injection (polyester urethane scaffold)	[78]
siRNA (target: PHD2)	Non-viral: micellar nanoparticles	Diabetic rat/ full-thickness	Injection (polythioketal urethane scaffold)	[79]
siRNA (target: GM3S)	Non-viral: nucleic acid gold nanoparticles	Diabetic mouse/ full-thickness wound	Topical	[80]
siRNA (target: MMP2)	Non-viral: LPEI nanoparticles	Diabetic mouse/ full-thickness wound	Topical	[81]
miRNA21	Non-viral: poloxamer gel	Mouse/ full-thickness wound	Topical	[82]
miRNA16, 20a, 21, 106a, 130a, 203	Non-viral: plasmid DNA	Diabetic rat/ full-thickness	Injection	[83]
miRNA29B	Non-viral: lipofectamine	Rat/full-thickness wound	Topical	[84]

2.6.1 Preclinical Studies

Vascular endothelial growth factor (VEGF) is important in wound healing because it promotes the early events in angiogenesis, especially endothelial cell migration and proliferation. VEGFs are produced by endothelial cells, keratinocytes, fibroblast, smooth muscle cells, platelets, neutrophils, and macrophages [97]. In a mouse wound healing model, Galeano et al. studied the effect of transduced VEGF on diabetes-associated wound healing by injecting adeno-associated virus (AAV) vector expressing VEGF intradermally into the wound edges. The results indicated that AAV vectors were highly efficient to transduce genes to the mouse skin. The VEGF levels in the wound specimens at day 7 and 14 were significantly higher compared to the control wounds, and the treatment improved wound healing in diabetic mice through the stimulation of angiogenesis, reepithelization, synthesis, and maturation of extracellular matrix [61]. In another mouse study, Galeano et al. used the same vector to treat partial-thickness thermal burns. Also in burns the gene transfer was

Table 2.2 Table summarizes the clinical studies (trial type, transgene, vector, transfer technique, and wound type and number) depicted in the text

Trial type	Transgene	Vector	Transfer technique	Wound type
Open-label [85]	VEGF	Non-viral: plasmid DNA	Injection	Ulcer ($n = 7$)
Open-label [86]	VEGF	Non-viral: plasmid DNA	Injection	Ulcer ($n = 6$)
Open-label [87]	HGF	Non-viral: plasmid DNA	Injection	Ulcer ($n = 11$)
Open-label [88]	HGF	Non-viral: plasmid DNA	Injection	Ulcer ($n = 10$)
Double-blinded, randomized, placebo-controlled [89]	HGF	Non-viral: plasmid DNA	Injection	Ulcer ($n = 50$)
Open-label [90]	HGF	Non-viral: plasmid DNA	Injection	Ulcer ($n = 9$)
Open-label [91–93]	PDGF-BB	Viral: adenovirus	Injection	Ulcer ($n = 15$)
Open-label [94]	PDGF-BB	Viral: adenovirus	Topical (collagen gel)	Ulcer ($n = 15$)
Double-blinded, randomized, placebo-controlled [95]	PDGF-BB	Viral: adenovirus	Topical (collagen gel)	Ulcer ($n = 72$)
Open-label [96]	FGF-1	Non-viral: plasmid DNA	Injection	Ulcer ($n = 9$)
Double-blinded, randomized, placebo-controlled [95]	FGF-1	Non-viral: plasmid DNA	Injection	Ulcer ($n = 107$)

The references to the studies are shown in parenthesis after the trial type

efficient, and the increased expression of VEGF enhanced burn wound repair by reducing inflammatory infiltration, stimulating dermal and epidermal regeneration, neovascularization, proliferation of fibroblasts, and the synthesis and maturation of extracellular matrix [62]. Kwon et al. delivered VEGF into diabetic mouse wounds using a non-viral gene delivery. They showed high transfection efficiency and enhanced wound healing by injecting VEGF encoding minicircle plasmid DNA in combination with an arginine-grafted cationic dendrimer. Using this non-viral method, the VEGF expression was first detected 4 days after the transfection, and the level gradually increased, doubling 12 days after administration. Treated diabetic wounds were generally healed within 6 days, whereas nontreated control wounds were over 40% open at day 6 [63].

Fibroblast growth factors (FGF) are produced by keratinocytes, fibroblasts, endothelial cells, smooth muscle cells, chondrocytes, and mast cells. They play a role in various stages during wound healing affecting granulation tissue formation, reepithelialization, and tissue remodeling [97–99]. Sun et al. used cationic liposomes to transfect FGF-1 into excisional wounds in diabetic mice. Their study showed that cells transfected with FGF-1 occupied up to 6% of the transectional area in the wound sites. The transfection was transient with a slow start, and the expression peaked at day 27 after administration. In addition, histology of the wounds revealed that the highest level of FGF expression was located in the normal tissue adjacent to the wounds rather than in the wound tissue. The treatment and increased FGF-1 expression in the diabetic mice wounds resulted in faster wound closure in comparison to control wounds [64]. Pereira et al. transferred FGF-7 (also known as keratinocyte growth factor, KGF) to acute rat wounds using cationic liposomes containing FGF-7 complementary DNA (cDNA). Their findings indicated that transfection of KFG into the wounds augmented wound healing through direct and indirect effects, which are not associated with pathologic scarring [65]. In another study, Escámez et al. compared the efficacy of different in vivo gene transfer strategies aimed at delivering FGF-7 to promote skin wound healing. FGF-7 gene was transferred to wounds by intradermal adenoviral injection, by adenoviral vector immobilized in a fibrin carrier or by FGF-7 adenoviral transduced human fibroblasts embedded in a fibrin matrix. All delivery systems achieved FGF-7 protein overproduction at the wound site resulting to faster reepithelialization [66].

During wound healing, *platelet-derived growth factors (PDGF)* are secreted by degranulating platelets, and they are present in the wound fluid. In addition, macrophages, vascular endothelium, fibroblasts, and keratinocytes produce PDGFs, which have a role in all the different phases of wound healing by promoting angiogenesis and granulation tissue formation, increasing collagen production, and stimulating cells to produce other growth factors [10, 97]. Eming et al. used retroviral vectors to transduce human keratinocytes ex vivo with a gene encoding human PDGF-AA. Transduction increased the level of keratinocytes PDGF secretion by 300-fold in vitro. Subsequently the cells were transplanted as epithelial sheets to athymic mice showing that PDGF-AA transplanted keratinocytes promoted growth and vascularization of the underlying dermis in comparison to not genetically modified keratinocyte sheets [67]. In another study, ischemic excisional rabbit ear wounds were treated with an adenovirus containing PDGF-B. The transduction resulted in a significant decrease in the epithelial gap in comparison to controls. The study also compared the efficacy of gene therapy and topical PDGF-BB protein treatment showing that the protein treatment had only modest effect on healing [68].

Growth factors that belong to *epidermal growth factor (EGF)* family stimulate cell growth, proliferation, and differentiation. They are secreted by platelets, macrophages, and fibroblasts, and in the skin and in wound healing, they stimulate especially keratinocytes in a paracrine manner [10, 97]. Andree et al. used particle-mediated DNA transfer to deliver human EGF to porcine partial-thickness wounds and showed 190-fold increase in the EGF concentrations in the wound fluids of the transfected wounds. The EGF transfected wounds healed 20% faster than the control wounds, and PCR results showed that the plasmid was present in the wound for at least 30 days after the gene transfer [69]. Rosenthal et al. transfected fibroblast ex vivo with EGF and transplanted the cells to murine full-thickness wounds. Their analysis showed increased EGF concentrations in transplanted wounds for 7 days [70].

Transforming growth factor beta (TGF-β) is produced by macrophages, fibroblasts, keratinocytes, and platelets. In wound healing, TGF-β is important in inflammation, angiogenesis, reepithelialization, and connective tissue regeneration [10,

97, 100]. In a study by Benn et al., a plasmid containing TGF-β was transfected using the particle-mediated DNA transfer to rat skin to determine whether overexpression of TGF-β at the site of skin incisions would result in a significant improvement in repair. Their results showed that TGF-β transfection resulted in a significant increase in tensile strength of healing rat tissue [71]. In another study Lee and Li used thermosensitive hydrogel containing plasmid TGF-β to treat full-thickness excisional diabetic mice wounds. Their results indicated significantly accelerated reepithelialization and increased cell proliferation in the wound bed treated with the plasmid TGF-β containing hydrogel. In addition, the accelerated reepithelialization was accompanied with enhanced collagen synthesis and more organized extracellular matrix deposition [72].

Hepatocyte growth factor/scatter factor (HGF) is one of the most versatile dermis-derived growth factors implicated in wound healing. It is secreted by mesenchymal cells and has a paracrine effect primarily on epithelial and endothelial cells being a powerful inducer of proliferation and migration of epithelial cells and a promoter of neoangiogenesis [101, 102]. Kunugiza et al. studied the efficacy of HGF gene transfer for wound healing in an impaired wound healing rat model. Gene transfer into the skin was conducted using a non-needle jet injector with a naked plasmid containing HGF, and 24 h after the transfection, increased HGF concentration was observed in the wounds. The results indicated that HGF gene transfer increased cellular proliferation of the epidermis and the number of microvessels in the granulation tissue and generally promoted wound healing in comparison to controls [73]. Also Ha et al. investigated the effect of HGF gene therapy on wound healing and on the prevention of scar formation. In a rabbit ear wound healing model, they transduced HGF using an adenoviral vector. Their results showed that application of HGF to the wounds significantly enhanced wound healing and reduced scarring. In addition, it was observed that the transduction of HFG cDNA into the wounds decreased the overexpression of TGF-β that plays an essential role in the progression of dermal fibrogenesis [74].

Insulin growth factors (IGFs) are synthesized by fibroblasts in the skin, and they stimulate mitogenesis, survival, and migration of keratinocytes in a paracrine manner. Many studies have reported a beneficial effect of exogenous IGF-1 on skin wound healing. IGF-1 levels have also been shown to significantly reduce in the skin of diabetic patients as well as in diabetic ulcers compared to nondiabetic wounds [103–105]. Hirsch et al. transfected keratinocytes ex vivo with cationic liposomes containing IGF-1 plasmid. The genetically modified cells were later transplanted into diabetic porcine wounds, and as result a 900-fold increase in IGF-1 expression was observed. The therapy also improved diabetic wound healing significantly [75]. Jeschke et al. studied the interaction between IGF-1 and other dermal and epidermal growth factors by administering IGF-1 to acute rat wounds as liposomal cDNA. Their results showed that IGF-1 gene transfer increased the concentrations of keratinocyte growth factor (KGF), FGF-1, PDGF, and type IV collagen that led to faster wound healing [76].

Gene therapy using *small interfering RNA (siRNA)* has been proposed as a potential treatment modality for many pathologies including wound healing.

siRNAs are small pieces of double-stranded RNA (20–25 base pairs in length) that bind to messenger RNA (mRNA), inhibiting the translation of the corresponding protein. This process is called RNA interference (RNAi) or siRNA silencing, and it can be translated into therapeutic applications. The biggest challenge in the therapeutic use of RNAi is the efficient delivery of siRNAs into target cells since naked siRNA cannot penetrate cell membranes and additionally has a very short half-life in vivo. Therefore, siRNA needs to be delivered to target tissue using carriers that can package, protect, and transport it into the cytoplasm of the target cell. In wound healing siRNAs can be utilized to knock down pro-inflammatory genes that are overexpressed in chronic wounds [60, 106]. Also well-known apoptotic derangements that occur in diabetic wounds can be targeted with this technique [107]. Thanik et al. developed an agarose matrix system to deliver a siRNA-liposomal transfection complex to cutaneous wounds in a murine model. Mitogen-activated protein kinase (Mapk-1) and lamin A/C were chosen as target genes due to their ubiquitous nature and high levels of expression during wound healing. The wound tissue was analyzed for target mRNA and protein suppression, and both in situ hybridization and Western blot analysis demonstrated significant reduction in Mapk-1 and lamin A/C expression in the wound bed at 14 and 21 days after wounding [77]. Nelson et al. delivered siRNA-carrying nanoparticles from hydrolytically biodegradable polyester urethane scaffolds and demonstrated in vivo knockdown of prolyl hydroxylase domain protein 2 (PHD2) with increased local angiogenesis in a subcutaneous mouse wound model [78]. In another study by the same group, the effect of PHD2 inhibition on wound healing was studied in diabetic rat wounds. They used porous polythioketal urethane scaffolds to locally deliver siRNA that inhibits the expression of PHD2. Their results showed that inhibition of PHD2 increased the expression of pro-growth genes leading to increased vasculature, cell proliferation, and tissue development in diabetic wounds [79]. Randeria et al. used siRNA-based spherical nucleic acid gold nanoparticle conjugates to silence ganglioside-monosialic acid 3 synthase (GM3S) that is overexpressed in diabetic mice and responsible for causing insulin resistance and impeding wound healing. Their results showed that the topical application of GM3S spherical nucleic acids to full-thickness wounds in diet-induced obese diabetic mice decreased local GM3S expression by >80% at the wound edge through a siRNA pathway and fully healed wounds within 12 day, whereas control wounds were only 50% closed [80]. Overexpression of matrix metalloproteinases (MMPs) has been depicted in diabetic wounds. Thus, in a diabetic murine wound model, Kim et al. used siRNA to inhibit MMP2 expression in wounds and studied the effect on wound healing. Their study indicated that diabetic ulcers treated with MMP2 siRNA showed faster wound closure, and the recovered tissue expressed a larger amount of cytokeratin along with a lower expression level of MMP2 compared to the control groups [81].

Another gene therapy modality is to transfer *microRNAs (miRNAs)*, which play key roles in the regulation of gene expression, to target tissues. miRNAs are small (21–23 base pairs in length) noncoding double-stranded RNAs that downregulate the expression of protein-coding genes. There are over 5000 miRNAs in the human genome that regulate over 30% of genomic RNAs, and several have been identified to influence wound healing process. Also in miRNA therapy, both viral and non-viral transfer techniques have been introduced, and development of effective and safety transfer techniques are key to translating this technology into patient care [59, 108–111]. Yang et al. studied the effect of miRNA21 on wound healing in a mouse model and showed that upregulation of miRNA21 at the wound edge promoted keratinocyte migration and boosted reepithelialization during healing. They further inhibited the upregulated expression of miRNA21 and demonstrated dramatically delayed reepithelialization [82]. Contrarily, Pastar et al. transfected a set of miRNAs (miRNA16, 20a, 21, 106a, 130a, and 203) into diabetic rat wounds and showed that the overexpression resulted as a loss of growth factor signaling, decreased reepithelialization, and granulation tissue formation that led to inhibition of healing [83]. Monaghan et al. used collagen scaffolds to transfect miRNA29B that is a posttranscriptional regulator of pro-fibrotic genes to full-thickness rat wounds and studied the effect of this structure on wound healing processes, specifically its impact on ECM remodeling following injury. Their results showed that the combination of collagen scaffold and miRNA29B reduced wound contraction and improved collagen III/I and MMP8/TIMP1 ratios in the wound bed in a dose-dependent manner [84, 112]. Li et al. showed that miRNA132 is highly upregulated during the inflammation phase of wound, mainly expressed in keratinocytes. They further found in a mouse wound healing model that miRNA132 blockade generated a severe inflammation reaction and deficiency in keratinocyte proliferation, therefore delaying wound healing. They concluded that miRNA132 is a critical regulator of skin wound healing facilitating the transition from the inflammatory phase to the proliferative phase [113].

2.6.2 Clinical Studies

In a phase I clinical trial of nine patients with non-healing ischemic ulcers ($n = 7$) and/or rest pain due to peripheral arterial disease, Baumgartner et al. transfected naked plasmids encoding human VEGF directly into the muscles of the ischemic limbs. Their objectives were to document the safety and feasibility of the gene transfer and analyze the potential therapeutic effect. Transient increase of VEGF expression was observed in serum levels 1–2 weeks after the gene transfer, and ischemic ulcers healed or markedly improved in four of seven limbs, including successful limb salvage in three patients. The only complication seen was transient lower-extremity edema, consistent with VEGF enhancement of vascular permeability [85]. Similarly, Kim et al. evaluated the safety and potential of intramuscular VEGF gene transfer. They treated nine patients suffering from severe peripheral arterial disease by intramuscular injection of a plasmid containing human VEGF to the limb. Six patients had non-healing ischemic ulcers, and after the treatment, ulcers healed or improved in four of the patients. Only three patients showed edema due to the VEGF enhancement of vascular permeability as a negative side effect [86].

Morishita et al. investigated the safety and efficacy of HGF plasmid DNA in patients with critical limb ischemia in an open-label clinical trial. Intramuscular injection of naked HGF plasmid DNA was performed to six patients with critical limb ischemia. A total of 11 ischemic ulcers were found in four patients. No severe complications caused by the HGF transfection were detected in any patients, and no apparent edema was observed in any patient throughout the trial. Twelve weeks after the transfection, 2 of 11 ulcers were completely healed, and 8 of 11 were significantly improved. Their study concluded that intramuscular injection of naked HGF plasmid is safe, feasible, and can achieve successful improvement of ischemic limbs [87]. In another open-label clinical study by the same group, ten patients with ischemic ulcers who were not candidates for revascularization and were unresponsive to conventional drug therapy were treated with HGF plasmid. The treatment consisted of eight injections of the plasmid (total HGF dose: 4 mg) into the calf muscles and/or distal thigh muscles of the ischemic limbs. Response to the treatment was evaluated from the reduction of ischemic ulcer size. The size of ischemic ulcers showed a decrease in 67% of the patients, and the ulcers healed completely in 56% of the patients after HGF gene therapy, and no major safety concerns were observed due to the therapy [88].

Powell et al. studied the effect of intramuscular HGF plasmid DNA injection in a double-blinded, randomized, placebo-controlled clinical trial at 20 centers with 104 patients suffering from critical limb ischemia. This study also indicated that the transfection of HGF plasmid was safe and well tolerated. However, no difference in wound healing or in other end points was observed among the groups [89]. In a phase I clinical trial, Yu et al. investigated the safety, tolerability, and preliminary efficacy of naked DNA therapy expressing two isoforms of HGF in critical limb ischemia patients. In total, 21 patients received increasing doses (Group 1, 4 mg; Group 2, 8 mg; Group 3, 12 mg; and Group 4, 16 mg) of HGF plasmid DNA, which were injected into the ischemic calf and/or thigh muscle at days 1 and 15. Enhanced wound healing was seen in in six out of nine patients having an ulcer at the last protocol visit (Group 1, 0/1; Group 2, 4/4; Group 3, 1/1; Group 4, 1/3). No severe complications due to the transfection were observed during the 3-month follow-up [90].

Margolis et al. have studied the safety and efficacy of PDGF-BB gene transfer to diabetic foot ulcers and venous leg ulcers. In their phase I clinical trial, PDGF-BB was transduced into the wounds by adenoviral vector. In total, 15 patients received a single dose of adenovirus encoding PDGF-BB. The dose was administered subcutaneously around ~3 cm of wound edge. The injection was well tolerated by all patients, and no signs of local wound infection, wound erythema, or cellulitis were observed as a consequence of the injection. All the subjects were tested for the presence of adenovirus and its ability to replicate at days 3, 7, 14, and 28. Adenovirus could not be cultured from any of the obtained specimens (blood, wound swabs, and bandage swabs), and additionally wound biopsies obtained at day 28 were negative for adenoviral DNA. All subjects showed improvement in their wounds over the 28-day follow-up. The amount of granulation tissue increased in all wounds, and a decrease in wound size was seen in 14 of 15 patients. In general, this phase I study demonstrated the initial safety, feasibility, and biologic plausibility of using adenoviral vectors to treat venous leg ulcer disease [91–93]. Mulder et al. incorporated a replication-defective adenoviral vector containing a human PDGF-BB into a biocompatible matrix (2.6% bovine collagen) and applied the gel topically onto wounds. In total, 15 patients with chronic non-healing foot ulcers took part into this open-label clinical trial. The results indicated that the treatment was well tolerated, and serum antibody titers to PDGF-B homodimer and collagen were negative, and adenoviral DNA was not detected in the blood. Rapid reduction in wound size was seen as a result of the treatment, and complete ulcer closure was observed in 10 patients by 3 months [94]. Blume et al. used the same collagen gel with and without PDGF encoding adenoviral vector to treat patients with diabetic neuropathic foot ulcers. Overall, 72 patients received adenoviral PDGF-B collagen gel, 33 patients received only gel, and 19 patients received standard of care. No adverse events related to the treatments were observed during the study. The ulcer closure incidences were 41% in gene therapy group, 45% in gel control group, and 31% in standard of care group. Therefore, no significant differences in wound healing were seen between the treatments during 4-week follow-up period [114].

In an open-label phase I clinical trial involving 51 patients, Comerota et al. investigated the effects and safety of intramuscular injections of naked plasmid DNA encoding FGF-1 on critical limb ischemia. The FGF-1 therapy was well tolerated, and no adverse effects were observed. Ulcer healing was evaluated in nine patients who entered the study with at least one preexisting ulcer. The results indicated that no patient had complete ulcer healing, defined as all ulcers healed and remaining healed until the end of the study. Three patients had complete healing of at least one ulcer that remained healed until the end of the study. Evaluation of the ulcer size showed progressive reduction over time with 68% of the surface area of all ulcers healed by the end of the follow-up (week 23) [96]. In the phase II clinical trial of the same FGF-1 therapy, 125 critical limb ischemia patients were randomized to non-viral FGF-1 ($n = 59$) or placebo ($n = 66$). groups. The gene transfer was well tolerated and resulted in a reduced risk of amputation. However, the use of non-viral FGF-1 did not demonstrate a statistically significant improvement in ulcer healing in comparison to the placebo group [95].

2.7 Challenges and Future Directions

A large number of preclinical studies in various animal models with multiple vectors and transfer techniques have been conducted over the years. Those studies have proven the feasibility of this treatment modality and shown some promising results and potential for clinical applications. In addition, they have indicated that gene therapy is a more efficient treatment modality than topical application of recombinant proteins for increasing the local concentration of a certain factor in a wound [22, 24, 115–117]. Nevertheless, translation of wound healing gene therapy strategies into clinical practice has been slow and challenging. Some clinical trials have been performed however with small patient groups and with relatively modest results leaving several issues concerning this therapy. The biggest challenge is still safety, and thus the possibility of immunological or toxic side effects, as well as tumor development, must be carefully studied in all cases. Another challenge is efficacy and durability of the treatment meaning that the used vector must be able to express the transgene at therapeutic levels for long enough time. Wound healing is a very complex process involving different steps and thousands of genes. Therefore, increasing the concentration of one gene is probably not sufficient enough to critically affect all the phases and complexities of wound repair. Finally, these therapies should be cost-effective and practical for doctors and nurses to use in order for them to become worthy options in the treatment of wounds [118, 119].

The future developments in gene therapy should include the transfer of multiple genes and further advances in siRNA and miRNA delivery. A few studies have demonstrated that the transfer of a combination of two growth factors has had more effect on wound healing than either growth factor alone [120, 121]. The RNAi by siRNA and miRNA delivery has a strong potential to be translated into therapeutic use once their roles in wound healing have been better understood. For example, suppressing overexpressed genes in the wounds can be used to treat non-healing chronic wounds or prevent hypertrophic scarring and keloid formation. In addition, this technology could be used to treat wound infection by targeting genes that are unique to bacteria and viruses thus inhibiting them from replicating [81, 108–111].

Also exploring the potential of emerging technologies such as clustered regularly interspaced short palindromic repeats (CRISPR)-associated nucleases such as Cas9 will be part of the future of wound healing gene therapy. CRISPR-Cas9 is a genome-editing tool that allows to edit genomes with unprecedented precision, efficiency, and flexibility, and it has been successfully used to generate disease models to study genetic diseases. CRISPR-Cas9 system consists of two elements that are guide sequence RNA (gRNA) and the Cas9 protein. By transferring the Cas9 protein with a gRNA into a cell, the cell's genome can be cut at a desired location, allowing genes to be removed and/or new ones added [122, 123]. This system could also potentially be very useful tool for wounds because CRISPR-Cas9 can both transiently and permanently genetically modify cells in the wound bed. In addition, this system can be used to modify the microenvironment of a chronic wound or prevent scarring by suppressing various genes [124].

References

1. Kruse CR, Nuutila K, Lee CC, Kiwanuka E, Singh M, Caterson EJ, Eriksson E, Sørensen JA. The external microenvironment of healing skin wounds. Wound Repair Regen. 2015;23(4):456–64. https://doi.org/10.1111/wrr.12303.
2. Nuutila K, Katayama S, Vuola J, Kankuri E. Human wound-healing research: issues and perspectives for studies using wide-scale analytic platforms. Adv Wound Care (New Rochelle). 2014;3(3):264–71.
3. Gurtner GC, Werner S, Barrandon Y, Longaker MT. Wound repair and regeneration. Nature. 2008;453(7193):314–21. https://doi.org/10.1038/nature07039.
4. Cohen S, Elliot GA. The stimulation of epidermal keratinization by a protein isolated from the submaxillary gland of the mouse. J Invest Dermatol. 1963;40:1–5.

5. Barrientos S, Brem H, Stojadinovic O, Tomic-Canic M. Clinical application of growth factors and cytokines in wound healing. Wound Repair Regen. 2014;22(5):569–78. https://doi.org/10.1111/wrr.12205.
6. Wieman TJ, Smiell JM, Su Y. Efficacy and safety of a topical gel formulation of recombinant human platelet-derived growth factor-BB (becaplermin) in patients with chronic neuropathic diabetic ulcers. A phase III randomized placebo-controlled double-blind study. Diabetes Care. 1998;21(5):822–7.
7. Bielefeld KA, Amini-Nik S, Alman BA. Cutaneous wound healing: recruiting developmental pathways for regeneration. Cell Mol Life Sci. 2013;70(12):2059–81. https://doi.org/10.1007/s00018-012-1152-9.
8. Yao F, Eriksson E. Gene therapy in wound repair and regeneration. Wound Repair Regen. 2000;8(6):443–51.
9. Eming SA, Martin P, Tomic-Canic M. Wound repair and regeneration: mechanisms, signaling, and translation. Sci Transl Med. 2014;6(265):265sr6. https://doi.org/10.1126/scitranslmed.3009337.
10. Kiwanuka E, Junker J, Eriksson E. Harnessing growth factors to influence wound healing. Clin Plast Surg. 2012;39(3):239–48. https://doi.org/10.1016/j.cps.2012.04.003.
11. Sun Y, Chen X, Xiao D. Tetracycline-inducible expression systems: new strategies and practices in the transgenic mouse modeling. Acta Biochim Biophys Sin (Shanghai). 2007;39(4):235–46.
12. Karzenowski D, Potter DW, Padidam M. Inducible control of transgene expression with ecdysone receptor: gene switches with high sensitivity, robust expression, and reduced size. BioTechniques. 2005;39(2):191–2, 194, 196.
13. Collins M, Thrasher A. Gene therapy: progress and predictions. Proc Biol Sci. 2015;282(1821):20143003. https://doi.org/10.1098/rspb.2014.3003.
14. Yin H, Kanasty RL, Eltoukhy AA, Vegas AJ, Dorkin JR, Anderson DG. Non-viral vectors for gene-based therapy. Nat Rev Genet. 2014;15(8):541–55. https://doi.org/10.1038/nrg3763.
15. Singer AJ, Clark RA. Cutaneous wound healing. N Engl J Med. 1999;341(10):738–46.
16. Guo S, Dipietro LA. Factors affecting wound healing. J Dent Res. 2010;89(3):219–29. https://doi.org/10.1177/0022034509359125.
17. Gill SE, Parks WC. Metalloproteinases and their inhibitors: regulators of wound healing. Int J Biochem Cell Biol. 2008;40(6–7):1334–47.
18. Widgerow AD. Chronic wound fluid--thinking outside the box. Wound Repair Regen. 2011;19(3):287–91. https://doi.org/10.1111/j.1524475X.2011.00683.x.
19. Muller M, Trocme C, Lardy B, Morel F, Halimi S, Benhamou PY. Matrix metalloproteinases and diabetic foot ulcers: the ratio of MMP-1 to TIMP-1 is a predictor of wound healing. Diabet Med. 2008;25(4):419–26. https://doi.org/10.1111/j.1464-5491.2008.02414.x.
20. Butzelaar L, Ulrich MM, Mink van der Molen AB, Niessen FB, Beelen RH. Currently known risk factors for hypertrophic skin scarring: a review. J Plast Reconstr Aesthet Surg. 2016;69(2):163–9. https://doi.org/10.1016/j.bjps.2015.11.015.
21. Fujio K, Komai T, Inoue M, Morita K, Okamura T, Yamamoto K. Revisiting the regulatory roles of the TGF-β family of cytokines. Autoimmun Rev. 2016;15(9):917–22. https://doi.org/10.1016/j.autrev.2016.07.007.
22. Branski LK, Pereira CT, Herndon DN, Jeschke MG. Gene therapy in wound healing: present status and future directions. Gene Ther. 2007;14(1):1–10.
23. Khavari PA, Rollman O, Vahlquist A. Cutaneous gene transfer for skin and systemic diseases. J Intern Med. 2002;252(1):1–10.
24. Branski LK, Gauglitz GG, Herndon DN, Jeschke MG. A review of gene and stem cell therapy in cutaneous wound healing. Burns. 2009;35(2):171–80. https://doi.org/10.1016/j.burns.2008.03.009.
25. Giatsidis G, Dalla Venezia E, Bassetto F. The role of gene therapy in regenerative surgery: updated insights. Plast Reconstr Surg. 2013;131(6):1425–35. https://doi.org/10.1097/PRS.0b013e31828bd153.

26. Appaiahgari MB, Vrati S. Adenoviruses as gene/vaccine delivery vectors: promises and pitfalls. Expert Opin Biol Ther. 2015;15(3):33751. https://doi.org/10.1517/14712598.2015.993374.
27. Kozarsky KF, Wilson JM. Gene therapy: adenovirus vectors. Curr Opin Genet Dev. 1993;3(3):499–503.
28. Petrie NC, Yao F, Eriksson E. Gene therapy in wound healing. Surg Clin North Am. 2003;83(3):597–616, vii.
29. Capasso C, Garofalo M, Hirvinen M, Cerullo V. The evolution of adenoviral vectors through genetic and chemical surface modifications. Virus. 2014;6(2):832–55. https://doi.org/10.3390/v6020832.
30. Gaj T, Epstein BE, Schaffer DV. Genome engineering using Adeno associated virus: basic and clinical research applications. Mol Ther. 2016;24(3):458–64. https://doi.org/10.1038/mt.2015.151.
31. Eming SA, Krieg T, Davidson JM. Gene transfer in tissue repair: status, challenges and future directions. Expert Opin Biol Ther. 2004;4(9):1373–86.
32. Hirsch T, Spielmann M, Yao F, Eriksson E. Gene therapy in cutaneous wound healing. Front Biosci. 2007;12:2507–18.
33. Burton EA, Fink DJ, Glorioso JC. Gene delivery using herpes simplex virus vectors. DNA Cell Biol. 2002;21(12):915–36.
34. Lachmann R. Herpes simplex virus-based vectors. Int J Exp Pathol. 2004;85(4):177–90.
35. Ramamoorth M, Narvekar A. Non viral vectors in gene therapy- an overview. J Clin Diagn Res. 2015;9(1):GE01–6. https://doi.org/10.7860/JCDR/2015/10443.5394.
36. Wang W, Li W, Ma N, Steinhoff G. Non-viral gene delivery methods. Curr Pharm Biotechnol. 2013;14(1):46–60.
37. Alsaggar M, Liu D. Physical methods for gene transfer. Adv Genet. 2015;89:1–24. https://doi.org/10.1016/bs.adgen.2014.10.001.
38. Nayerossadat N, Maedeh T, Ali PA. Viral and nonviral delivery systems for gene delivery. Adv Biomed Res. 2012;1:27. https://doi.org/10.4103/2277-9175.98152.
39. Gibot L, Rols MP. Gene transfer by pulsed electric field is highly promising in cutaneous wound healing. Expert Opin Biol Ther. 2016;16(1):67–77. https://doi.org/10.1517/14712598.2016.1098615.
40. Wells DJ. Gene therapy progress and prospects: electroporation and other physical methods. Gene Ther. 2004;11(18):1363–9.
41. Hoeller D, Petrie N, Yao F, Eriksson E. Gene therapy in soft tissue reconstruction. Cells Tissues Organs. 2002;172(2):118–25.
42. Davidson JM, Krieg T, Eming SA. Particle-mediated gene therapy of wounds. Wound Repair Regen. 2000;8(6):452–9.
43. Eriksson E, Yao F, Svensjö T, Winkler T, Slama J, Macklin MD, Andree C, McGregor M, Hinshaw V, Swain WF. In vivo gene transfer to skin and wound by microseeding. J Surg Res. 1998;78(2):85–91.
44. Jeschke MG, Sandmann G, Finnerty CC, Herndon DN, Pereira CT, Schubert T, Klein D. The structure and composition of liposomes can affect skin regeneration, morphology and growth factor expression in acute wounds. Gene Ther. 2005;12(23):1718–24.
45. Bleiziffer O, Eriksson E, Yao F, Horch RE, Kneser U. Gene transfer strategies in tissue engineering. J Cell Mol Med. 2007;11(2):206–23.
46. Felgner PL, Ringold GM. Cationic liposome-mediated transfection. Nature. 1989;337:387–8.
47. Hengge UR, et al. Cytokine gene expression in epidermis with biological effects following injection of naked DNA. Nat Genet. 1995;10:161–6.
48. Dang JM, Leong KW. Natural polymers for gene delivery and tissue engineering. Adv Drug Deliv Rev. 2006;58(4):487–99.
49. Cam C, Segura T. Matrix-based gene delivery for tissue repair. Curr Opin Biotechnol. 2013;24(5):855–63. https://doi.org/10.1016/j.copbio.2013.04.007.
50. Chandler LA, Doukas J, Gonzalez AM, Hoganson DK, Gu DL, Ma C, Nesbit M, Crombleholme TM, Herlyn M, Sosnowski BA, Pierce GF. FGF2-targeted adenovirus encoding platelet-derived growth factor-B enhances de novo tissue formation. Mol Ther. 2000;2(2):153–60.

51. Lee RJ, Springer ML, Blanco-Bose WE, Shaw R, Ursell PC, Blau HM. VEGF gene delivery to myocardium: deleterious effects of unregulated expression. Circulation. 2000;102(8):898–901.
52. Toniatti C, Bujard H, Cortese R, Ciliberto G. Gene therapy progress and prospects: transcription regulatory systems. Gene Ther. 2004;11(8):649–57.
53. Yao F, Pomahac B, Visovatti S, Chen M, Johnson S, Augustinova H, Svensjo T, Eriksson E. Systemic and localized reversible regulation of transgene expression by tetracycline with tetR-mediated transcription repression switch. J Surg Res. 2007;138(2):267–74.
54. Gower RM, Shea LD. Biomaterial scaffolds for controlled, localized gene delivery of regenerative factors. Adv Wound Care (New Rochelle). 2013;2(3):100–6.
55. Shepard JA, Wesson PJ, Wang CE, Stevans AC, Holland SJ, Shikanov A, Grzybowski BA, Shea LD. Gene therapy vectors with enhanced transfection based on hydrogels modified with affinity peptides. Biomaterials. 2011;32:5092.
56. Eming SA, Krieg T, Davidson JM. Gene therapy and wound healing. Clin Dermatol. 2007;25(1):79–92.
57. Nuutila K, Siltanen A, Peura M, Bizik J, Kaartinen I, Kuokkanen H, Nieminen T, Harjula A, Aarnio P, Vuola J, Kankuri E. Human skin transcriptome during superficial cutaneous wound healing. Wound Repair Regen. 2012;20(6):830–9. https://doi.org/10.1111/j.1524-475X.2012.00831.x.
58. Broughton G, Janis JE, Attinger CE. The basic science of wound healing. Plast Reconstr Surg. 2006;117(7 Suppl):12S–34S.
59. Peng B, Chen Y, Leong KW. MicroRNA delivery for regenerative medicine. Adv Drug Deliv Rev. 2015;88:108–22. https://doi.org/10.1016/j.addr.2015.05.014.
60. Nelson CE, Gupta MK, Adolph EJ, Guelcher SA, Duvall CL. siRNA delivery from an injectable scaffold for wound therapy. Adv Wound Care (New Rochelle). 2013;2(3):93–9.
61. Galeano M, Deodato B, Altavilla D, Cucinotta D, Arsic N, Marini H, Torre V, Giacca M, Squadrito F. Adeno-associated viral vector-mediated human vascular endothelial growth factor gene transfer stimulates angiogenesis and wound healing in the genetically diabetic mouse. Diabetologia. 2003;46(4):546–55.
62. Galeano M, Deodato B, Altavilla D, Squadrito G, Seminara P, Marini H, Stagno d'Alcontres F, Colonna M, Calò M, Lo Cascio P, Torre V, Giacca M, Venuti FS, Squadrito F. Effect of recombinant adeno-associated virus vector-mediated vascular endothelial growth factor gene transfer on wound healing after burn injury. Crit Care Med. 2003;31(4):1017–25.
63. Kwon MJ, An S, Choi S, Nam K, Jung HS, Yoon CS, Ko JH, Jun HJ, Kim TK, Jung SJ, Park JH, Lee Y, Park JS. Effective healing of diabetic skin wounds by using nonviral gene therapy based on minicircle vascular endothelial growth factor DNA and a cationic dendrimer. J Gene Med. 2012;14(4):272–8. https://doi.org/10.1002/jgm.2618.
64. Sun L, Xu L, Chang H, Henry FA, Miller RM, Harmon JM, Nielsen TB. Transfection with aFGF cDNA improves wound healing. J Invest Dermatol. 1997;108(3):313–8.
65. Pereira CT, Herndon DN, Rocker R, Jeschke MG. Liposomal gene transfer of keratinocyte growth factor improves wound healing by altering growth factor and collagen expression. J Surg Res. 2007;139(2):222–8.
66. Escámez MJ, Carretero M, García M, Martínez-Santamaría L, Mirones I, Duarte B, Holguín A, García E, García V, Meana A, Jorcano JL, Larcher F, Del Río M. Assessment of optimal virus-mediated growth factor gene delivery for human cutaneous wound healing enhancement. J Invest Dermatol. 2008;128(6):1565–75. https://doi.org/10.1038/sj.jid.5701217.
67. Eming SA, Lee J, Snow RG, Tompkins RG, Yarmush ML, Morgan JR. Genetically modified human epidermis overexpressing PDGF-A directs the development of a cellular and vascular connective tissue stroma when transplanted to athymic mice--implications for the use of genetically modified keratinocytes to modulate dermal regeneration. J Invest Dermatol. 1995;105(6):756–63.
68. Liechty KW, Nesbit M, Herlyn M, Radu A, Adzick NS, Crombleholme TM. Adenoviral-mediated overexpression of platelet-derived growth factor-B corrects ischemic impaired wound healing. J Invest Dermatol. 1999;113(3):375–83.

69. Andree C, Swain WF, Page CP, Macklin MD, Slama J, Hatzis D, Eriksson E. In vivo transfer and expression of a human epidermal growth factor gene accelerates wound repair. Proc Natl Acad Sci U S A. 1994;91(25):1218892.
70. Rosenthal FM, Cao L, Tanczos E, Kopp J, Andree C, Stark GB, Mertelsmann R, Kulmburg P. Paracrine stimulation of keratinocytes in vitro and continuous delivery of epidermal growth factor to wounds in vivo by genetically modified fibroblasts transfected with a novel chimeric construct. In Vivo. 1997;11(3):201–8.
71. Benn SI, Whitsitt JS, Broadley KN, Nanney LB, Perkins D, He L, Patel M, Morgan JR, Swain WF, Davidson JM. Particle-mediated gene transfer with transforming growth factor-beta1 cDNAs enhances wound repair in rat skin. J Clin Invest. 1996;98(12):2894–902.
72. Lee PY, Li Z, Huang L. Thermosensitive hydrogel as a Tgf-beta1 gene delivery vehicle enhances diabetic wound healing. Pharm Res. 2003;20(12):1995–2000.
73. Kunugiza Y, Tomita N, Taniyama Y, Tomita T, Osako MK, Tamai K, et al. Acceleration of wound healing by combined gene transfer of hepatocyte growth factor and prostacyclin synthase with Shima Jet. Gene Ther. 2006;13:1143–52.
74. Ha X, Li Y, Lao M, Yuan B, Wu CT. Effect of human hepatocyte growth factor on promoting wound healing and preventing scar formation by adenovirus-mediated gene transfer. Chin Med J. 2003;116:1029–33.
75. Hirsch T, Spielmann M, Velander P, Zuhaili B, Bleiziffer O, Fossum M, Steinstraesser L, Yao F, Eriksson E. Insulin-like growth factor-1 gene therapy and cell transplantation in diabetic wounds. J Gene Med. 2008;10:1247–52.
76. Jeschke MG, Schubert T, Krickhahn M, Polykandriotis E, Klein D, Perez-Polo JR, et al. Interaction of exogenous liposomal insulin-like growth factor-I cDNA gene transfer with growth factors on collagen expression in acute wounds. Wound Repair Regen. 2005;13:269–77.
77. Thanik VD, Greives MR, Lerman OZ, Seiser N, Dec W, Chang CC, Warren SM, Levine JP, Saadeh PB. Topical matrix-based siRNA silences local gene expression in a murine wound model. Gene Ther. 2007;14(17):1305–8.
78. Nelson CE, Kim AJ, Adolph EJ, Gupta MK, Yu F, Hocking KM, Davidson JM, Guelcher SA, Duvall CL. Tunable delivery of siRNA from a biodegradable scaffold to promote angiogenesis in vivo. Adv Mater. 2014;26(4):607–14, 506. https://doi.org/10.1002/adma.201303520.
79. Martin JR, Nelson CE, Gupta MK, Yu F, Sarett SM, Hocking KM, Pollins AC, Nanney LB, Davidson JM, Guelcher SA, Duvall CL. Local delivery of PHD2 siRNA from ROS-degradable scaffolds to promote diabetic wound healing. Adv Healthc Mater. 2016;5(21):2751–7. https://doi.org/10.1002/adhm.201600820.
80. Randeria PS, Seeger MA, Wang XQ, Wilson H, Shipp D, Mirkin CA, Paller AS. siRNA-based spherical nucleic acids reverse impaired wound healing in diabetic mice by ganglioside GM3 synthase knockdown. Proc Natl Acad Sci U S A. 2015;112(18):5573–8. https://doi.org/10.1073/pnas.1505951112.
81. Kim HS, Son YJ, Yoo HS. Clustering siRNA conjugates for MMP-responsive therapeutics in chronic wounds of diabetic animals. Nanoscale. 2016;8(27):13236–44. https://doi.org/10.1039/c6nr01551d.
82. Yang X, Wang J, Guo SL, Fan KJ, Li J, Wang YL, Teng Y. MiR-21 promotes keratinocyte migration and re-epithelialization during wound healing. Int J Biol Sci. 2011;7:685–90.
83. Pastar I, Khan AA, Stojadinovic O, Lebrun EA, Medina MC, Brem H, Kirsner RS, Jimenez JJ, Leslie C, Tomic-Canic M. Induction of specific microRNAs inhibits cutaneous wound healing. J Biol Chem. 2012;287:29324–35.
84. Monaghan M, Browne S, Schenke-Layland K, Pandit A. A collagen-based scaffold delivering exogenous microRNA-29b to modulate extracellular matrix remodeling. Mol Ther. 2014;22:786–96.
85. Baumgartner I, Pieczek A, Manor O, Blair R, Kearney M, Walsh K, Isner JM. Constitutive expression of phVEGF165 after intramuscular gene transfer promotes collateral vessel development in patients with critical limb ischemia. Circulation. 1998;97:1114–23.

86. Kim HJ, Jang SY, Park JL, Byun J, Kim DL, Do YS, Kim JM, Kim S, Kim BM, Kim WB, Kim DK. Vascular endothelial growth factor-induced angiogenic gene therapy in patients with peripheral artery disease. Exp Mol Med. 2004;36:336–44.
87. Morishita R, Aoki M, Hashiya N, Makino H, Yamasaki K, Azuma J, Sawa Y, Matsuda H, Kaneda Y, Ogihara T. Safety evaluation of clinical gene therapy using hepatocyte growth factor to treat peripheral arterial disease. Hypertension. 2004;44:203–9.
88. Shigematsu H, Yasuda K, Sasajima T, Takano T, Miyata T, Ohta T, Tanemoto K, Obitsu Y, Iwai T, Ozaki S, Ogihara T, Morishita R, HGF Study Group. Transfection of human HGF plasmid DNA improves limb salvage in Buerger's disease patients with critical limb ischemia. Int Angiol. 2011;30(2):140–9.
89. Powell RJ, Simons M, Mendelsohn FO, Daniel G, Henry TD, Koga M, Morishita R, Annex BH. Results of a double-blind, placebo-controlled study to assess the safety of intramuscular injection of hepatocyte growth factor plasmid to improve limb perfusion in patients with critical limb ischemia. Circulation. 2008;118(1):58–65. https://doi.org/10.1161/CIRCULATIONAHA.107.727347.
90. Gu Y, Zhang J, Guo L, Cui S, Li X, Ding D, Kim JM, Ho SH, Hahn W, Kim S. A phase I clinical study of naked DNA expressing two isoforms of hepatocyte growth factor to treat patients with critical limb ischemia. J Gene Med. 2011;13(11):602–10. https://doi.org/10.1002/jgm.1614.
91. Margolis DJ, Crombleholme T, Herlyn M. Clinical protocol: phase I trial to evaluate the safety of H5.020CMV.PDGF-B for the treatment of a diabetic insensate foot ulcer. Wound Repair Regen. 2000;8:480–93.
92. Margolis DJ, Cromblehome T, Herlyn M, Cross P, Weinberg L, Filip J, Propert K. Clinical protocol. Phase I trial to evaluate the safety of H5.020CMV.PDGF-b and limb compression bandage for the treatment of venous leg ulcer: trial A. Hum Gene Ther. 2004;15:1003–19.
93. Margolis DJ, Morris LM, Papadopoulos M, Weinberg L, Filip JC, Lang SA, Vaikunth SS, Crombleholme TM. Phase I study of H5.020CMV.PDGF-beta to treat venous leg ulcer disease. Mol Ther. 2009;17(10):1822–9. https://doi.org/10.1038/mt.2009.169.
94. Mulder G, Tallis AJ, Marshall VT, Mozingo D, Phillips L, Pierce GF, Chandler LA, Sosnowski BK. Treatment of nonhealing diabetic foot ulcers with a platelet-derived growth factor gene-activated matrix (GAM501): results of a phase 1/2 trial. Wound Repair Regen. 2009;17(6):772–9. https://doi.org/10.1111/j.1524-475X.2009.00541.x.
95. Nikol S, Baumgartner I, Van Belle E, Diehm C, Visoná A, Capogrossi MC, Ferreira-Maldent N, Gallino A, Wyatt MG, Wijesinghe LD, Fusari M, Stephan D, Emmerich J, Pompilio G, Vermassen F, Pham E, Grek V, Coleman M, Meyer F, TALISMAN 201 Investigators. Therapeutic angiogenesis with intramuscular NV1FGF improves amputation-free survival in patients with critical limb ischemia. Mol Ther. 2008;16(5):972–8. https://doi.org/10.1038/mt.2008.33.
96. Comerota AJ, Throm RC, Miller KA, Henry T, Chronos N, Laird J, Sequeira R, Kent CK, Bacchetta M, Goldman C, Salenius JP, Schmieder FA, Pilsudski R. Naked plasmid DNA encoding fibroblast growth factor type 1 for the treatment of end-stage unreconstructible lower extremity ischemia: preliminary results of a phase I trial. J Vasc Surg. 2002;35(5):930–6.
97. Barrientos S, Stojadinovic O, Golinko MS, Brem H, Tomic-Canic M. Growth factors and cytokines in wound healing. Wound Repair Regen. 2008;16(5):585–601. https://doi.org/10.1111/j.1524-475X.2008.00410.x.
98. Bennett SP, Griffiths GD, Schor AM, Leese GP, Schor SL. Growth factors in the treatment of diabetic foot ulcers. Br J Surg. 2003;90:133–46.
99. Powers CJ, McLeskey SW, Wellstein A. Fibroblast growth factors, their receptors and signaling. Endocr Relat Cancer. 2000;7:165–97.
100. Ramirez H, Patel SB, Pastar I. The role of TGFβ signaling in wound epithelialization. Adv Wound Care (New Rochelle). 2014;3(7):482–91.
101. Peura M, Bizik J, Salmenperä P, Noro A, Korhonen M, Pätilä T, Vento A, Vaheri A, Alitalo R, Vuola J, Harjula A, Kankuri E. Bone marrow mesenchymal stem cells undergo nemosis and induce keratinocyte wound healing utilizing the HGF/c-Met/PI3K pathway. Wound Repair Regen. 2009;17(4):569–77. https://doi.org/10.1111/j.1524-475X.2009.00507.x.

102. Chmielowiec J, Borowiak M, Morkel M, Stradal T, Munz B, Werner S, Wehland J, Birchmeier C, Birchmeier W. c-Met is essential for wound healing in the skin. J Cell Biol. 2007;177(1):151–62.
103. Seeger MA, Paller AS. The roles of growth factors in keratinocyte migration. Adv Wound Care (New Rochelle). 2015;4(4):213–24.
104. Werner S, Grose R. Regulation of wound healing by growth factors and cytokines. Physiol Rev. 2003;83:835–70.
105. Aydin F, Kaya A, Karapinar L, Kumbaraci M, Imerci A, Karapinar H, Karakuzu C, Incesu M. IGF-1 increases with hyperbaric oxygen therapy and promotes wound healing in diabetic foot ulcers. J Diabetes Res. 2013;2013:567834. https://doi.org/10.1155/2013/567834.
106. Aljuffali IA, Lin YK, Fang JY. Noninvasive approach for enhancing small interfering RNA delivery percutaneously. Expert Opin Drug Deliv. 2016;13(2):265–80. https://doi.org/10.1517/17425247.2016.1121988.
107. Darby IA, Bisucci T, Hewitson TD, MacLellan DG. Apoptosis is increased in a model of diabetes-impaired wound healing in genetically diabetic mice. Int J Biochem Cell Biol. 1997;29:191–200.
108. Beavers KR, Nelson CE, Duvall CL. MiRNA inhibition in tissue engineering and regenerative medicine. Adv Drug Deliv Rev. 2015;88:123–37. https://doi.org/10.1016/j.addr.2014.12.006.
109. Banerjee J, Sen CK. MicroRNAs in skin and wound healing. Methods Mol Biol. 2013;936:343–56.
110. Broderick JA, Zamore PD. MicroRNA therapeutics. Gene Ther. 2011;18(12):1104–10. https://doi.org/10.1038/gt.2011.50.
111. Fahs F, Bi X, Yu FS, Zhou L, Mi QS. New insights into microRNAs in skin wound healing. IUBMB Life. 2015;67(12):889–96. https://doi.org/10.1002/iub.1449.
112. Maurer B, Stanczyk J, Jüngel A, Akhmetshina A, Trenkmann M, Brock M, Kowal-Bielecka O, Gay RE, Michel BA, Distler JH, Gay S, Distler O. MicroRNA-29, a key regulator of collagen expression in systemic sclerosis. Arthritis Rheum. 2010;62(6):1733–43. https://doi.org/10.1002/art.27443.
113. Li D, Wang A, Liu X, Meisgen F, Grünler J, Botusan IR, Narayanan S, Erikci E, Li X, Blomqvist L, Du L, Pivarcsi A, Sonkoly E, Chowdhury K, Catrina SB, Ståhle M, Landén NX. MicroRNA-132 enhances transition from inflammation to proliferation during wound healing. J Clin Invest. 2015;125(8):3008–26. https://doi.org/10.1172/JCI79052.
114. Blume P, Driver VR, Tallis AJ, Kirsner RS, Kroeker R, Payne WG, Wali S, Marston W, Dove C, Engler RL, Chandler LA, Sosnowski BK. Formulated collagen gel accelerates healing rate immediately after application in patients with diabetic neuropathic foot ulcers. Wound Repair Regen. 2011;19(3):302–8. https://doi.org/10.1111/j.1524-475X.2011.00669.x.
115. Steed DL. Modifying the wound healing response with exogenous growth factors. Clin Plast Surg. 1998;25:397–405.
116. Robson MC, Mustoe TA, Hunt TK. The future of recombinant growth factors in wound healing. Am J Surg. 1998;176:80S–2S.
117. Robson MC, Smith PD. Topical use of growth factors to enhance healing. In: Falanga V, editor. Cutaneous wound healing. London: Taylor & Francis; 2001. p. 379–98.
118. Li S, Huang L. Nonviral gene therapy: promises and challenges. Gene Ther. 2000;7(1):31–4.
119. Whittam AJ, Maan ZN, Duscher D, Wong VW, Barrera JA, Januszyk M, Gurtner GC. Challenges and opportunities in drug delivery for wound healing. Adv Wound Care (New Rochelle). 2016;5(2):79–88.
120. Lynch SE, Nixon JC, Colvin RB, Antoniades HN. Role of platelet-derived growth factor in wound healing: synergistic effects with other growth factors. Proc Natl Acad Sci U S A. 1987;84:7696–700.
121. Sprugel KH, McPherson JM, Clowes AW, Ross R. Effects of growth factors in vivo. I. Cell ingrowth into porous subcutaneous chambers. Am J Pathol. 1987;129:601–13.
122. Eid A, Mahfouz MM. Genome editing: the road of CRISPR/Cas9 from bench to clinic. Exp Mol Med. 2016;48(10):e265. https://doi.org/10.1038/emm.2016.111.

123. Wang M, Glass ZA, Xu Q. Non-viral delivery of genome-editing nucleases for gene therapy. Gene Ther. 2016;24(3):144. https://doi.org/10.1038/gt.2016.72.
124. Sessions JW, Armstrong DG, Hope S, Jensen BD. A review of genetic engineering biotechnologies for enhanced chronic wound healing. Exp Dermatol. 2016;26(2):179–85. https://doi.org/10.1111/exd.13185.

3 Gene Therapy in Pathologic Scars

Chenyu Huang, Longwei Liu, Zhifeng You, Yanan Du, and Rei Ogawa

3.1 Introduction

Pathological scars are common fibroproliferative disorders that occur in the largest organ of the integumentary system. They include hypertrophic scars (HSs) and keloids, whose key histological characteristics are the accumulation of fibroblasts and collagen fibers. HSs differ from keloids in that they undergo chronic clinical contracture and develop pathological dermal nodes. By contrast, keloids exhibit clinically significant invasion growth that extends into the neighboring healthy skin; it is often erythematous and pruritic and contains pathological hyalinized collagens [1, 2]. The incidence of HS is as high as 40–70% after surgery and up to 91% after burn injuries [3–5], while the prevalence of keloids ranges from 0.09% in England to 16% in Zaire [6]. The etiology of these abnormal scars remains unclear, although there is evidence that genetics [7], local skin tension [8], sebum secretion [9], and neurogenic inflammation [10] contribute. The multifactorial nature of the etiology of these scars means that the therapeutic strategies for them are generally nonspecific and palliative and that recurrences are rather common.

These issues have led to strong interest in the development of novel interventions that can prevent, reduce, or even reverse pathological scar formation and/or

C. Huang
Department of Dermatology, Beijing Tsinghua Changgung Hospital, School of Clinical Medicine, Tsinghua University, Beijing, China

L. Liu · Z. You · Y. Du
Department of Biomedical Engineering, School of Medicine, Tsinghua University, Beijing, China
e-mail: llw13@mails.tsinghua.edu.cn; duyanan@tsinghua.edu.cn

R. Ogawa (✉)
Department of Plastic, Reconstructive and Aesthetic Surgery, Nippon Medical School, Tokyo, Japan
e-mail: r.ogawa@nms.ac.jp

G. Giatsidis (ed.), *Gene Therapy in Reconstructive and Regenerative Surgery*,
https://doi.org/10.1007/978-3-319-78957-6_3

progression. In particular, there is currently a marked and growing focus on genetic interventions for HSs and keloids such as short-interfering RNAs (siRNA). This is largely because gene therapy may have a number of potential advantages compared to the classical pharmacological therapies that are currently being used to manage pathological scars, namely, glucocorticoid and transforming growth factor (TGF)-β protein inhibitor. The advantages of gene therapies are as follows. (1) Gene therapies are more specific and thereby can prevent keloid recurrence. Glucocorticoids (e.g., dexamethasone) are currently the gold standard pharmacological therapy for keloid management. They ameliorate keloids by reducing collagen I/III expression at the mRNA (not the protein) level [11]. However, they also associate with a substantial rate of keloid recurrence [12]. This may reflect the fact that they also stimulate connective tissue growth factor (CTGF) expression [11]. (2) Gene therapies have more long-lived effects**.** Another pharmacological therapy for keloids consists of specific monoclonal antibodies that inhibit a target protein. An example is the anti-TGF-β antibody [13, 14]. This antibody has similar inhibitory effects as gene therapies that target the same molecule, such as siRNA against TGF-β receptor; both therapies inhibit cell proliferation and the gene expression of the factors that operate downstream of TGF-β. However, anti-TGF-β antibody specifically targets the TGF-β proteins, which have a short half-life in vivo. Therefore, to maintain the effect, repeated applications of the antibody are needed. By contrast, a single application of TGF-β receptor-inhibiting siRNA reduces TGF-β protein concentrations as effectively as the antibody, and repeated applications are not needed to sustain this effect, at least in vitro [15]. (3) Gene therapy may be cheaper. This is exemplified by the previous example: a single dose of TGF-β receptor mRNA-targeting gene therapy is likely to be less expensive than protein-targeting therapy, which requires multiple doses to maintain the response [16].

3.2 Basic Gene Therapy Techniques

The genetic therapies for cutaneous fibrosis can be divided according to whether the delivery approach does or does not involve viruses [17–19]. (1) Non-viral methods. Gene transfer via non-viral DNA or RNA could avoid the pathogenic effects of viral methods. It has been shown that direct intradermal injection of DNA plasmids into wound sites results in site-specific gene expression targeted to dermal and subdermal layers, which can still be detectable 8 weeks after the injection [20]. Physical methods such as the gene gun and micro-seeding can also be used to deliver genes [17]. The gene gun involves particle-mediated (e.g., gold particles) delivery of the target gene in vivo [21, 22]. For example, when the epidermal growth factor gene is delivered into partial-thickness excisional wounds in pig skin by using a gene gun, epidermal growth factor protein is overexpressed for 10 days and associates with accelerated reepithelialization and wound closure [23]. Similarly, when murine fibrosarcomas are transfected with FMS-like tyrosine kinase-3 ligand by using a gene gun, the growth of the tumor is suppressed [24]. This indicates the potential of this molecule as a scar treatment. Micro-seeding involves the delivery of a plasmid DNA solution to the target cells of the skin by using oscillating solid microneedles that are driven by a modified tattooing device. It is more efficient than either single injections or particle-mediated gene transfer [25]. (2) Viral methods**.** Viral techniques include the use of retroviruses, adeno-associated viruses, and adenoviruses. Viral methods deliver the gene more efficiently than non-viral technologies. However, all viral approaches also associate with cellular toxicity, immunogenicity, and unwanted genetic mutations, which significantly limit their use in human gene therapy [18, 19]. Each viral system also associates with various strengths and limitations. *Retroviruses* can cause the target gene to stably and efficiently integrate into the host's chromosomal DNA without inducing rearrangements. However, this system is significantly hampered by its limited packing and transferring size (normally <6 kb). Moreover, the integration of the transfected gene is dependent on cell division, which also limits the usefulness of this method [26]. In addition, it remains possible that the retrovirus method could activate a cellular proto-oncogene after integration, which is a significant safety issue that should not be ignored [27, 28]. *Adeno-associated virus* has an advantage over the retrovirus system: it not only integrates into the cellular genome; it can infect both dividing and nondividing cells. Studies in animal models have shown that gene transfer using adeno-associated virus results in long-term transgene expression without inducing immune responses or other toxicities [29]. However, this system is limited by the small transgene size (up to 4.7 kb), the complex preparation of adeno-associated virus (e.g., wild-type adenovirus must be inactivated), and potential safety issues due to the nature of the gene integration mechanism. By contrast, *adenoviruses* not only infect both dividing and nondividing cells; they rarely integrate into the host genome because the viruses induce strong inflammatory responses in the transfection site that eventually destroy the transduced cells. This means that this system is unlikely to cause unwanted oncogene activation. It also means that this system can only induce short-term transgene expression. While this could be seen as a significant limitation, this feature has made adenoviruses particularly suitable for cases where transient gene expression is needed, such as wounded skin [26]. Adenoviruses can also carry exogenous DNA up to 7.5 kb in length and can be easily manipulated to produce high viral titers, which results in further infection of nondividing cells with high efficiency [17].

Both injections and viral approaches have been used to deliver the most widely applied gene therapy for scars, namely, microRNA (miRNA)-based therapy. MiRNAs are 18–25-nucleotide single-stranded noncoding RNA molecules that alter gene expression by binding to target mRNA, thereby inducing the degradation of the mRNA or preventing protein translation. In skin fibrosis, they can play either a pro- or antifibrotic role. MiRNA-based scar therapies thus aim to downregulate the pro-fibrotic miRNAs and/or upregulate the antifibrotic miRNAs [30]. For example, treatments that reduce miRNA-21 levels also downregulate the epidermal-mesenchymal transition that characterizes fibrosis [31, 32] In addition, treatments that elevate miRNA-146a levels inhibit Smad4 protein expression, which in turn downregulates TGF-β signaling and its pro-fibrotic effects, including extracellular matrix (ECM) deposition [33]. MiRNA levels can be reduced by using anti-miRNAs, which are synthetic complementary oligonucleotides that bind to the target miRNAs and thereby inhibit miRNA activity. Other competitive inhibitors of miRNAs are the so-called sponges, erasers, and masks [34]. Sponges contain several tandem sites that bind to the miRNA seed region (nucleotides 2–7); this allows the sponge to inhibit all miRNAs that contain a common seed region. MiRNA erasers are similar to sponges except that they employ two copies of the perfectly complementary antisense sequence of miRNA. Masks do not bind to the target miRNA; rather, they bind to the mRNA sequence that is targeted by the miRNA, and thereby prevent the miRNA/mRNA association [34].

The key features of pathological scarring are rampant fibroblast and myofibroblast proliferation, accumulation of collagens in the ECM, and altered regulation of growth factors. Thus, we propose that the current gene therapy strategies for HSs and keloids can be classified on the basis of whether they target cellular behavior, the production of collagens in the ECM, or growth factors/cytokines, as follows. (1) Gene therapies that target cellular behavior. These therapies include those that promote apoptosis or inhibit the proliferation of dermal fibroblasts. Besides, epithelial to mesenchymal transition or transformation (EMT) is to be inhibited. The therapies also include those that enhance epidermal keratinocyte motility and vascular endothelial cell proliferation: these approaches accelerate the reepithelialization of the wound, thereby preventing the chronic course of wound healing that characterizes pathological scarring. (2) Gene therapies that prevent ECM accumulation. Collagen I is the main ECM ingredient in HSs and keloids. Its accumulation is due to an imbalance between the matrix metalloproteinases (MMPs) and their inhibitors, namely, tissue inhibitors of metalloproteinases (TIMPs). Thus, gene therapies that target the MMP/TIMP system could alter the abnormal accumulation of collagen I in pathological scars. This either inhibits collagen I synthesis or promotes collagen I degradation. (3) Gene therapies that target growth factors/cytokines. TGF-β has a well-known pro-fibrotic role. Genetic therapies have been used to downregulate TGF-β by decreasing its activation or by inhibiting TGF-β-signaling pathway molecules such as TGF-β receptors and Smads molecules.

3.2.1 Gene Therapies that Alter the Balance Between Cellular Proliferation and Apoptosis

Fibroblast apoptosis has been induced by double-suicide gene therapy. The double-suicide gene CDglyTK consists of *Escherichia coli* cytosine deaminase (CD) linked to herpes simplex virus thymidine kinase (TK) via a polyglycine chain. When cells are transfected with CDglyTK, they and nearby cells become highly susceptible to 5-fluorocytosine and ganciclovir [35]. When keloid fibroblasts were treated with adenovirus bearing CDglyTK (Ad-CMV-CDglyTK), the cells exhibited inhibited growth due to cellular apoptosis, as shown by their decreased expression of Bcl-2 and increased expression of BAX [36].

Fibroblast apoptosis has also been induced by lentiviral vector-mediated expression of tumor necrosis factor receptor (TNF)-associated death domain protein (TRADD). Keloid tissues exhibit decreased expression of TRADD, which is supported by the reduced apoptotic activity in keloid fibroblasts [37]. Indeed, when HS fibroblasts were transfected with the PLVX-TRADD-EGFP lentivirus and treated with 10 ng/mL TNF-α, they expressed TRADD-GFP-FLAG fusion protein, their intracellular TRADD levels normalized, their proliferation decreased, and they exhibited G2/M arrest. These changes are associated with decreased collagen I secretion [38]. Thus, the recombinant TRADD-expressing lentivirus may be useful for reducing HS formation and/or progression.

Fibroblast proliferation can also be reduced by manipulating miRNA-21 levels. Keloid tissues express 6.87-fold higher levels of miRNA-21 than normal skin samples [39]. A study on cancer cells, which also overexpress miRNA-21, showed that miRNA-21 increases cellular proliferation and invasion by suppressing phosphatase and tensin homolog (PTEN) protein levels. It was also shown that MiR-21 suppresses PTEN by directly binding to its 3′ untranslated region [40]. Like many cancer cells, keloids also have low PTEN levels relative to adjacent normal skin. When keloid fibroblasts were transfected with mimicking or inhibiting oligonucleotides that were complementary to miRNA-21, their PTEN levels correlated negatively with their cellular miRNA-21 levels. Moreover, miRNA-21 inhibition in keloid cells decreased their phosphorylated Akt levels, which is suggestive of decreased proliferation. Furthermore, normal fibroblasts exhibited increased proliferation when they were transfected with miRNA-21 mimics [41]. Thus, gene therapy targeting miRNA-21 may help to reduce abnormal scar development or progression.

3.2.2 Gene Therapies that Alter the Balance Between Collagen Synthesis and Degradation

The strong accumulation of collagens, especially collagen I, in HSs and keloids means that many gene therapies that aim to inhibit collagen I synthesis or promoting its degradation have been considered. Gene therapies that regulate the interactions between collagens and interfibrillar matrix proteins have also been proposed.

One approach to reducing collagen I production focuses on miRNA-29b. It was shown recently that transfection with this miRNA potently downregulates the collagen I synthesis of primary skin fibroblasts at both the messenger and protein levels. Moreover, skin scars express significantly lower levels of miR-29b and these miRNA-29b levels correlate inversely with collagen I subunit levels [42].

Another approach to reducing collagen synthesis in keloids is to inhibit plasminogen activator inhibitor (PAI). Keloid fibroblasts overexpress PAI-1. Notably, when PAI-1 expression is upregulated in normal fibroblasts by PAI-1-expressing adenovirus, the cells exhibit a two to fourfold increase in extracellular collagen accumulation in vitro. Moreover, PAI-1-targeting siRNAs and lentivirus that expressed a short hairpin RNA (shRNA)-containing microRNA (shRNAmir) sequence for PAI-1 reduced the collagen synthesis of both normal and keloid fibroblasts [43]. Similarly, Syed et al. showed recently that when keloid explants were organ-cultured on collagen gel matrix at the air-liquid interface of serum-free medium and then transfected with 250 nM PAI-1-targeting siRNA, the explants exhibited significantly more shrinkage and significantly less collagen accumulation than the untreated explants [11]. Thus, PAI-1-targeting interventions such as siRNA and shRNAmir-expressing lentivirus transfection may help to prevent keloid development or progression.

Yet another way of suppressing the collagen production of abnormal scars is to block heat shock protein (HSP) 47. HSP47 is a specific chaperone for collagens that assists in the folding, assembly, and transport of these molecules. Thus, it is required for the full maturation of collagens. Consequently, HSP47 expression is largely seen in the endoplasmic reticulum of collagen-producing cells [44, 45]. Keloids and HSs express HSP47 mRNA at significantly higher levels than normal skin controls [46]. Moreover, when keloid fibroblasts were transduced with plasmids containing HSP47-shRNA in vitro, they exhibited a marked decrease in not only HSP47 mRNA and protein levels but also collagen synthesis and secretion [47]. This was also observed in vivo: when subcutaneously implanted keloid tissues on nude mice were transfected with the same plasmids, they exhibited greater shrinkage and less collagen than the control tissues [48]. In accordance with these observations, when full-thickness skin wounds (1 cm diameter) were induced on the backs of mice and then injected with an adenovirus expressing HSP47 siRNA, they exhibited significantly less collagen I expression, and the formation of scars was substantially reduced [49]. Thus, HSP47, shRNA, or siRNA may suppress the excessive collagen accumulation in keloids in an active, specific, and efficient manner.

Scar tissue is mechanically weaker than uninjured tissue because the normal highly ordered large-diameter collagen fibrils are not well-formed. The proteoglycan decorin plays an important role in collagen fibrillogenesis [50], as shown by the well-known fact that decorin associates with type-VI collagen [51] and interacts with fibronectin, a major non-collagenous matrix protein [52]. The latter observation is particularly important because fibronectin plays a key role in soft tissue healing [53]. In particular, the binding of decorin to fibronectin may help regulate the interfibrillar bonding of collagens. Nakamura et al. speculated that downregulating decorin expression after wounding would improve the mechanical properties of scars [50]. To test this, they encapsulated decorin antisense oligonucleotides with a liposome that was complexed to the fusion proteins of hemagglutinating virus of Japan; the fusion of this liposome with cell membrane introduces the DNA in the liposome directly into the cytoplasm. Moreover, when the liposome was co-introduced with a DNA-binding nuclear protein, it ensured the efficient nuclear translocation of the DNA to the nucleus [54, 55]. Indeed, Nakamura et al. showed that when rabbit ligament wounds were treated only once with liposomes containing decorin antisense oligonucleotides, their expression of decorin was significantly suppressed, and larger diameter collagen fibrils were observed 4 weeks after treatment. Moreover, the lower the expression of decorin mRNA, the larger the collagen fibril diameters were. Strength and creep tests also showed that these changes improved the strength of the scars and reduced their elongation [50]. Thus, manipulating collagen fibrillogenesis during the repair of soft connective tissue by decorin gene therapy could improve the scars that arise after wounding.

Another way to downregulate the collagen accumulation that characterizes abnormal scars is to induce collagen degradation and remodeling by targeting MMPs and their inhibitors, the TIMPs. TIMP-2 is usually downregulated in keloid-derived fibroblasts relative to normal fibroblasts from the same individuals [56]. When keloid fibroblasts were transduced with a lentiviral vector that expressed siRNA targeting TIMP-1 or TIMP-2, the complexing of these TIMPs to MMP-1 was suppressed, and MMP-2 activity was upregulated. Moreover, the siTMP-1-expressing keloid fibroblasts exhibited increased degradation of collagen I in the supernatant; however, this was not observed in the siTIMP-2-expressing cells. This suggests that siTIMP-1RNA could treat keloids by increasing collagen bundle degradation [57]. In addition, Lee et al. showed that when keloid-derived fibroblasts were transduced with a decorin-expressing adenovirus, their MMP-1 and MMP-3 mRNA levels rose sharply, and their collagen I and III mRNA levels dropped. Similarly, immunohistochemical analysis of keloid spheroids transduced with the same decorin-expressing adenovirus showed that they expressed much less collagen I and III than the control spheroids [58]. Thus, this approach could help reduce keloid development/progression, as well as postoperative recurrence. Similarly, when Li et al. transduced normal human foreskin fibroblasts with the human indoleamine 2,3-dioxygenase (IDO) gene, their MMP-1 expression rose significantly. IDO is a rate-limiting enzyme that converts L-tryptophan into kynurenine. When the fibroblasts were treated with kynurenine, their MMP-1 and MMP-3 expression increased markedly. It was found that this effect was mediated by activation of the MEK (mitogen-activated protein kinase (MAPK)/extracellular signal-regulated kinase (ERK) kinase)-ERK1/2 MAPK signaling pathway. Li et al. then showed that when fibrotic wounds on rabbit ears were treated with topical application of a cream containing 50 ug kynurenine, their scar formation improved significantly [59]. Thus, gene therapy with IDO or topical application of kynurenine may improve abnormal scars.

3.2.3 Gene Therapies that Alter the Balance Between Pro-Fibrotic and Anti-Fibrotic Growth Factors/Cytokines

TGF-β signaling is well documented to mediate wound healing and fibrosis [60, 61]. TGF-β signals through a heteromeric receptor complex of type I and II receptor serine/threonine kinases [62]; the signal is then propagated downstream through Smads [63]. Thus, miRNA-based therapies that target various TGF-β pathway molecules such as latency-associated peptide (LAP)-TGFβ3, TGF-β receptor 1, and Smad4 may be useful gene therapies for HSs and keloids.

Waddington et al. developed a lentiviral vector that expressed a mutant form of murine TGF-β3 that was secreted as the stable LAP-associated form. The mutation did not affect the activation of the cytokine by various mechanisms, including integrin, but greatly enhanced its bioavailability. This in turn augmented the ability of TGF-β3 to antagonize the pro-fibrotic TGF-β1 molecule. When full-thickness cutaneous incision wounds on mouse flanks were intradermally injected with this vector, the reepithelialization density and fibroblast/myofibroblast transdifferentiation in the wound area decreased significantly. Thus, local intradermal transduction with this TGF-β pathway-based gene therapy reduced scar tissue formation [64].

When human HS fibroblasts were transduced with TGF-β receptor 1 (TGFBR1) siRNA in vitro, their cell proliferation was suppressed. This effect is associated with decreased Smad2/Smad3 phosphorylation, reduced CTGF mRNA expression, and lower protein and mRNA expression of various ECM molecules. The same siRNA also had in vivo effects on abnormal scars, namely, model chronic HS scars that were generated by inducing a large full-thickness skin wound on rabbit ears. When the granulated tissues in the late proliferative and early remodeling phases of wound healing were injected with the TGFBR1 siRNA, ECM deposition and scar contracture were reduced. These in vivo effects are associated with decreased CTGF and α-smooth muscle actin mRNA expression [15].

Tan et al. developed an adenovirus that expressed mutant Smad4 (Ad-ΔM4), which antagonizes wild-type Smad4. When they transfected mouse fibroblasts with Ad-ΔM4 in vitro, it effectively blocked the TGF-β signaling pathway; collagen I and III mRNA and protein expression were downregulated, and fibroblast proliferation was inhibited. Furthermore, when Ad-ΔM4 was applied externally to a full-thickness incision wound deep in the panniculus muscle on the rat dorsal skin, the wound healed with much less scarring than the control wounds. Thus, such externally applied gene therapy may be an effective and convenient way to inhibit scar formation [65].

Gene therapies that target growth factors other than TGF-β or its signaling pathway molecules may also be feasible and effective methods to prevent or reduce pathological scarring. An example is focal adhesion kinase (FAK). FAK mediates fibroblast migration by activating integrin through downstream of the Janus Kinase (JAK) pathway; it also contributes to the signal transduction that occurs in response to TGF-β. When HS fibroblasts are transduced with FAK siRNA in vitro, their integrin-α, TGF-β, and α-smooth muscle actin mRNA expression is reduced by 49.7%, 59.0%, and 49.1%, respectively. This associated with a strong decrease in HS fibroblast proliferation [66]. Moreover, when an adenovirus expressing hepatic growth factor (HGF) was exogenously applied to an excessive scar in rabbits that resembled human HSs in terms of clinical and histological features, the established skin scars dramatically flattened, with thinner dermal layer and diminished amount of collagen I. And the HGF-expressing adenovirus improved the scars potentially by inhibiting TGF-β1 expression, according to Ha et al. [67].

Another example of the effectiveness of gene therapies targeting non-TGF-β molecules is the study by Sisco et al. They first found that the HS-like scars of rabbit models had very high levels of CTGF mRNA. When they injected full-thickness wounds on rabbit ears with the antisense oligonucleotide of CTGF, the collagen I expression decreased [68].

These observations together indicate that gene therapies that employ non-viral or viral methods and that aim to induce keloid fibroblast apoptosis, inhibit collagen accumulation, or block TGF-β signaling pathways or other growth factors have therapeutic potential for abnormal scar management. However, the gene therapy studies that have been reported to date mainly examined the effect of the therapies on target cells in vitro. The comparative shortage of in vivo evidence at this stage largely reflects our limited understanding of the etiology of keloids and the lack of animal models for this pathological scar. Fortunately, however, recent years have seen considerable and increasing research into how gene therapies can reshape the balance between insufficient and excessive healing and thereby improve wound healing. To achieve the ultimate goal of gene therapy, namely, converting the scar-forming wound healing process into a scarless or reduced scar healing process, future research efforts should focus on two directions: (1) the identification of gene-specific promoters [69] or enhancers [70] that direct the expression of transfected genes into specific target cells would help as the target modification of specific cells in the skin; and (2) the use of tools such as the CRISPR-Cas9 system to achieve precise gene modification of the target tissues; this would facilitate precise targeted changes in the genome of the skin when it is combined with traditional gene delivery methods [71, 72]. A better understanding of the current progress in gene therapies for pathological scars will almost certainly lead to the development of novel interventions that can prevent, reduce, or even reverse the formation and/or progression of HSs or keloids.

References

1. Lee JY, Yang CC, Chao SC, Wong TW. Histopathological differential diagnosis of keloid and hypertrophic scar. Am J Dermatopathol. 2004;26(5):379–84.
2. Huang C, Akaishi S, Hyakusoku H, Ogawa R. Are keloid and hypertrophic scar different forms of the same disorder? A fibroproliferative skin disorder hypothesis based on keloid findings. Int Wound J. 2014;11(5):517–22.
3. Deitch EA, Wheelahan TM, Rose MP, Clothier J, Cotter J. Hypertrophic burn scars: analysis of variables. J Trauma. 1983;23(10):895–8.
4. Lewis WH, Sun KK. Hypertrophic scar: a genetic hypothesis. Burns. 1990;16(3):176–8.
5. Gauglitz GG, Korting HC, Pavicic T, Ruzicka T, Jeschke MG. Hypertrophic scarring and keloids:pathomechanisms and current and emerging treatment strategies. Mol Med. 2011;17(1–2):113–25.
6. Louw L. Keloids in rural black South Africans. Part 1: general overview and essential fatty acid hypotheses for keloid formation and prevention. Prostaglandins Leukot Essent Fatty Acids. 2000;63(5):237–45.
7. Marneros AG, Norris JE, Watanabe S, Reichenberger E, Olsen BR. Genome scans provide evidence for keloid susceptibility loci on chromosomes 2q23 and 7p11. J Invest Dermatol. 2004;122(5):1126–32.
8. Ogawa R. Keloid and hypertrophic scarring may result from a mechanoreceptor or mechanosensitive nociceptor disorder. Med Hypotheses. 2008;71(4):493–500.
9. Fong EP, Bay BH. Keloids – the sebum hypothesis revisited. Med Hypotheses. 2002;58(4):264–9.
10. Akaishi S, Ogawa R, Hyakusoku H. Keloid and hypertrophic scar: neurogenic inflammation hypotheses. Med Hypotheses. 2008;71(1):32–8.
11. Syed F, Bagabir RA, Paus R, Bayat A. Ex vivo evaluation of antifibrotic compounds in skin scarring: EGCG and silencing of PAI-1 independently inhibit growth and induce keloid shrinkage. Lab Investig. 2013;93(8):946–60.

12. Juckett G, Hartman-Adams H. Management of keloids and hypertrophic scars. Am Fam Physician. 2009;80(3):253–60.
13. Lu L, Saulis AS, Liu WR, Roy NK, Chao JD, Ledbetter S, Mustoe TA. The temporal effects of anti-TGF-beta1, 2, and 3 monoclonal antibody on wound healing and hypertrophic scar formation. J Am Coll Surg. 2005;201(3):391–7.
14. Mead AL, Wong TT, Cordeiro MF, Anderson IK, Khaw PT. Evaluation of anti-TGF-beta2 antibody as a new postoperative anti-scarring agent in glaucoma surgery. Invest Ophthalmol Vis Sci. 2003;44(8):3394–401.
15. Wang YW, Liou NH, Cherng JH, Chang SJ, Ma KH, Fu E, Liu JC, Dai NT. siRNA-targeting transforming growth factor-β type I receptor reduces wound scarring and extracellular matrix deposition of scar tissue. J Invest Dermatol. 2014;134(7):2016–25.
16. Liu W, Chua C, Wu X, Wang D, Ying D, Cui L, Cao Y. Inhibiting scar formation in rat wounds by adenovirus-mediated overexpression of truncated TGF-beta receptor II. Plast Reconstr Surg. 2005;115(3):860–70.
17. Liu W, Cao Y, Longaker MT. Gene therapy of scarring: a lesson learned from fetal scarless wound healing. Yonsei Med J. 2001;42(6):634–45.
18. Eriksson E. Gene transfer in wound healing. Adv Skin Wound Care. 2000;13(2 Suppl):20–2.
19. Mulligan RC. The basic science of gene therapy. Science. 1993;260(5110):926–32.
20. Meuli M, Liu Y, Liggitt D, Kashani-Sabet M, Knauer S, Meuli-Simmen C, Harrison MR, Adzick NS, Heath TD, Debs RJ. Efficient gene expression in skin wound sites following local plasmid injection. J Invest Dermatol. 2001;116(1):131–5.
21. Yang CH, Shen SC, Lee JC, Wu PC, Hsueh SF, Lu CY, Meng CT, Hong HS, Yang LC. Seeing the gene therapy: application of gene gun technique to transfect and decolour pigmented rat skin with human agouti signalling protein cDNA. Gene Ther. 2004;11(13):1033–9.
22. Davidson JM, Eming SA, Dasgupta J. Particle-mediated gene therapy of wounds. Methods Mol Med. 2003;78:433–52.
23. Andree C, Swain WF, Page CP, Macklin MD, Slama J, Hatzis D, Eriksson E. In vivo transfer and expression of a human epidermal growth factor gene accelerates wound repair. Proc Natl Acad Sci U S A. 1994;91(25):12188–92.
24. Abe A, Furumoto H, Yoshida K, Kato T, Saijo Y, Irahara M. Gene gun-mediated skin transfection with FL gene suppresses the growth of murine fibrosarcoma. J Med Investig. 2011;58(1-2):39–45.
25. Eriksson E, Yao F, Svensjö T, Winkler T, Slama J, Macklin MD, Andree C, McGregor M, Hinshaw V, Swain WF. In vivo gene transfer to skin and wound by microseeding. J Surg Res. 1998;78(2):85–91.
26. Eming SA, Krieg T, Davidson JM. Gene therapy and wound healing. Clin Dermatol. 2007;25(1):79–92.
27. Yi Y, Noh MJ, Lee KH. Current advances in retroviral gene therapy. Curr Gene Ther. 2011;11(3):218–28.
28. Yi Y, Hahm SH, Lee KH. Retroviral gene therapy: safety issues and possible solutions. Curr Gene Ther. 2005;5(1):25–35.
29. Daya S, Berns KI. Gene therapy using adeno-associated virus vectors. Clin Microbiol Rev. 2008;21(4):583–93.
30. Babalola O, Mamalis A, Lev-Tov H, Jagdeo J. The role of microRNAs in skin fibrosis. Arch Dermatol Res. 2013;305(9):763–76.
31. Han M, Wang Y, Liu M, Bi X, Bao J, Zeng N, Zhu Z, Mo Z, Wu C, Chen X. MiR-21 regulates epithelial-mesenchymal transition phenotype and hypoxia-inducible factor-1α expression in third-sphere forming breast cancer stem cell-like cells. Cancer Sci. 2012;103(6):1058–64.
32. Kumarswamy R, Volkmann I, Thum T. Regulation and function of miRNA-21 in health and disease. RNA Biol. 2011;8(5):706–13.
33. Liu Z, Lu CL, Cui LP, Hu YL, Yu Q, Jiang Y, Ma T, Jiao DK, Wang D, Jia CY. MicroRNA-146a modulates TGF-β1-induced phenotypic differentiation in human dermal fibroblasts by targeting SMAD4. Arch Dermatol Res. 2012;304(3):195–202.

34. Fasanaro P, Greco S, Ivan M, Capogrossi MC, Martelli F. microRNA: emerging therapeutic targets in acute ischemic diseases. Pharmacol Ther. 2010;125(1):92–104.
35. Xia K, Liang D, Tang A, Feng Y, Zhang J, Pan Q, Long Z, Dai H, Cai F, Wu L, Zhao S, Chen Z, Xia J. A novel fusion suicide gene yeast CDglyTK plays a role in radio-gene therapy of nasopharyngeal carcinoma. Cancer Gene Ther. 2004;11(12):790–6.
36. Xu B, Liu ZZ, Zhu GY, Yang JF, Zhao JP, Wang JC, Cai JL. Efficacy of recombinant adenovirus-mediated double suicide gene therapy in human keloid fibroblasts. Clin Exp Dermatol. 2008;33(3):322–8.
37. Sayah DN, Soo C, Shaw WW, Watson J, Messadi D, Longaker MT, Zhang X, Ting K. Downregulation of apoptosis-related genes in keloid tissues. J Surg Res. 1999;87(2):209–16.
38. Yuan Y, Peng G, Kang X, Liu Y, Dai J, Wu H. Effects of lentiviral vector-mediated TRADD expression on the inhibition of hypertrophic scar formation. Exp Biol Med (Maywood). 2014;239(12):1557–66.
39. Liu Y, Yang D, Xiao Z, Zhang M. miRNA expression profiles in keloid tissue and corresponding normal skin tissue. Aesthet Plast Surg. 2012;36(1):193–201.
40. Zhang JG, Wang JJ, Zhao F, Liu Q, Jiang K, Yang GH. MicroRNA-21 (miR-21) represses tumor suppressor PTEN and promotes growth and invasion in non-small cell lung cancer (NSCLC). Clin Chim Acta. 2010;411(11–12):846–52.
41. Liu Y, Wang X, Yang D, Xiao Z, Chen X. MicroRNA-21 affects proliferation and apoptosis by regulating expression of PTEN in human keloid fibroblasts. Plast Reconstr Surg. 2014;134(4):561e–73e.
42. Cheng J, Wang Y, Wang D, Wu Y. Identification of collagen 1 as a post-transcriptional target of miR-29b in skin fibroblasts: therapeutic implication for scar reduction. Am J Med Sci. 2013;346(2):98–103.
43. Tuan TL, Hwu P, Ho W, Yiu P, Chang R, Wysocki A, Benya PD. Adenoviral overexpression and small interfering RNA suppression demonstrate that plasminogen activator inhibitor-1 produces elevated collagen accumulation in normal and keloid fibroblasts. Am J Pathol. 2008;173(5):1311–25.
44. Taguchi T, Razzaque MS. The collagen-specific molecular chaperone HSP47: is there a role in fibrosis? Trends Mol Med. 2007;13(2):45–53.
45. Koide T, Asada S, Takahara Y, Nishikawa Y, Nagata K, Kitagawa K. Specific recognition of the collagen triple helix by chaperone HSP47: minimal structural requirement and spatial molecular orientation. J Biol Chem. 2006;281(6):3432–8.
46. Chen J, Cen Y. Detection of expression of heat shock protein 47 mRNA in pathological scar tissue by using real-time fluorescent quantitative RT-PCR. Zhongguo Xiu Fu Chong Jian Wai Ke Za Zhi. 2005;19(8):613–6.
47. Chen JJ, Zhao S, Cen Y, Liu XX, Yu R, Wu DM. Effect of heat shock protein 47 on collagen accumulation in keloid fibroblast cells. Br J Dermatol. 2007;156(6):1188–95.
48. Chen JJ, Jin PS, Zhao S, Cen Y, Liu Y, Xu XW, Duan WQ, Wang HS. Effect of heat shock protein 47 on collagen synthesis of keloid in vivo. ANZ J Surg. 2011;81(6):425–30.
49. Wang Z, Li L. Adenovirus-mediated RNA interference against collagen-specific molecular chaperone 47-KDa heat shock protein suppresses scar formation on mouse wounds. Cell Biol Int. 2008;32(5):484–93.
50. Nakamura N, Hart DA, Boorman RS, Kaneda Y, Shrive NG, Marchuk LL, Shino K, Ochi T, Frank CB. Decorin antisense gene therapy improves functional healing of early rabbit ligament scar with enhanced collagen fibrillogenesis in vivo. J Orthop Res. 2000;18(4):517–23.
51. Takahashi T, Cho HI, Kublin CL, Cintron C. Keratan sulfate and dermatan sulfate proteoglycans associate with type VI collagen in fetal rabbit cornea. J Histochem Cytochem. 1993;41(10):1447–57.
52. Schmidt G, Robenek H, Harrach B, Glössl J, Nolte V, Hörmann H, Richter H, Kresse H. Interaction of small dermatan sulfate proteoglycan from fibroblasts with fibronectin. J Cell Biol. 1987;104(6):1683–91.

53. Welch MP, Odland GF, Clark RA. Temporal relationships of F-actin bundle formation, collagen and fibronectin matrix assembly, and fibronectin receptor expression to wound contraction. J Cell Biol. 1990;110(1):133–45.
54. Dzau VJ, Mann MJ, Morishita R, Kaneda Y. Fusigenic viral liposome for gene therapy in cardiovascular diseases. Proc Natl Acad Sci U S A. 1996;93(21):11421–5.
55. Kaneda Y, Morishita R, Tomita N. Increased expression of DNA cointroduced with nuclear protein in adult rat liver. J Mol Med (Berl). 1995;73(6):289–97.
56. Dohi T, Miyake K, Aoki M, Ogawa R, Akaishi S, Shimada T, Okada T, Hyakusoku H. Tissue inhibitor of metalloproteinase-2 suppresses collagen synthesis in cultured keloid fibroblasts. Plast Reconstr Surg Glob Open. 2015;3(9):e520.
57. Aoki M, Miyake K, Ogawa R, Dohi T, Akaishi S, Hyakusoku H, Shimada T. siRNA knockdown of tissue inhibitor of metalloproteinase-1 in keloid fibroblasts leads to degradation of collagen type I. J Invest Dermatol. 2014;134(3):818–26.
58. Lee WJ, Ahn HM, Roh H, Na Y, Choi IK, Lee JH, Kim YO, Lew DH, Yun CO. Decorin-expressing adenovirus decreases collagen synthesis and upregulates MMP expression in keloid fibroblasts and keloid spheroids. Exp Dermatol. 2015;24(8):591–7.
59. Li Y, Kilani RT, Rahmani-Neishaboor E, Jalili RB, Ghahary A. Kynurenine increases matrix metalloproteinase-1 and -3 expression in cultured dermal fibroblasts and improves scarring in vivo. J Invest Dermatol. 2014;134(3):643–50.
60. Ashcroft GS, Yang X, Glick AB, Weinstein M, Letterio JL, Mizel DE, Anzano M, Greenwell-Wild T, Wahl SM, Deng C, Roberts AB. Mice lacking Smad3 show accelerated wound healing and an impaired local inflammatory response. Nat Cell Biol. 1999;1(5):260–6.
61. Flanders KC. Smad3 as a mediator of the fibrotic response. Int J Exp Pathol. 2004;85(2):47–64.
62. Massagué J, Attisano L, Wrana JL. The TGF-beta family and its composite receptors. Trends Cell Biol. 1994;4(5):172–8.
63. Massagué J, Wotton D. Transcriptional control by the TGF-beta/Smad signaling system. EMBO J. 2000;19(8):1745–54.
64. Waddington SN, Crossley R, Sheard V, Howe SJ, Buckley SM, Coughlan L, Gilham DE, Hawkins RE, McKay TR. Gene delivery of a mutant TGFβ3 reduces markers of scar tissue formation after cutaneous wounding. Mol Ther. 2010;18(12):2104–11.
65. Tan WQ, Gao ZJ, Xu JH, Yao HP. Inhibiting scar formation in vitro and in vivo by adenovirus-mediated mutant Smad4: a preliminary report. Exp Dermatol. 2011;20(2):119–24.
66. Chen R, Zhang Z, Xue Z, Wang L, Fu M, Lu Y, Bai L, Zhang D, Fan Z. Focal adhesion kinase (FAK) siRNA inhibits human hypertrophic scar by suppressing integrin α, TGF-β and α-SMA. Cell Biol Int. 2014;38(7):803–8.
67. Ha X, Yuan B, Li Y, Lao M, Wu Z. Gene therapy for pathological scar with hepatocyte growth factor mediated by recombinant adenovirus vector. Sci China C Life Sci. 2003;46(3):320–7.
68. Sisco M, Kryger ZB, O'Shaughnessy KD, Kim PS, Schultz GS, Ding XZ, Roy NK, Dean NM, Mustoe TA. Antisense inhibition of connective tissue growth factor (CTGF/CCN2) mRNA limits hypertrophic scarring without affecting wound healing in vivo. Wound Repair Regen. 2008;16(5):661–73.
69. Chen X, Scapa J, Liu D, Godbey W. Cancer-specific promoters for expression-targeted gene therapy: ran, brms1, and mcm5. J Gene Med. 2016;18(7):89–101. https://doi.org/10.1002/jgm.2882.
70. Lu Q, Ganjawala TH, Ivanova E, Cheng JG, Troilo D, Pan ZH. AAV-mediated transduction and targeting of retinal bipolar cells with improved mGluR6 promoters in rodents and primates. Gene Ther. 2016;23(8–9):680–9. https://doi.org/10.1038/gt.2016.42.
71. Cong L, Ran FA, Cox D, Lin S, Barretto R, Habib N, Hsu PD, Wu X, Jiang W, Marraffini LA, Zhang F. Multiplex genome engineering using CRISPR/Cas systems. Science. 2013;339(6121):819–23.
72. Ran FA, Hsu PD, Wright J, Agarwala V, Scott DA, Zhang F. Genome engineering using the CRISPR-Cas9 system. Nat Protoc. 2013;8(11):2281–308.

Gene Therapy in Skeletal Muscle Repair and Regeneration

4

Adriana C. Panayi*, Shane M. Lince*, and Indranil Sinha

4.1 Introduction

Gene therapy (GT), or the altering of genetic material within cells to modify protein structure and quantity, is an emerging field in skeletal muscle regeneration (SMR) and maintenance [1–6]. GT techniques are acquiring increasingly significant functions in approaching various surgical problems [7, 8]. A key insight necessary for maximizing the value from the great deal already known on the regulation of SMR is that every regulatory product is potentially amenable to targeting through multiple GT techniques. This chapter will focus on current GT techniques used in muscle discussing, in particular, potential benefits for Plastic and Reconstructive Surgery (PRS).

Skeletal muscle is the largest tissue in the human body, accounting for 40–50% of total mass, and plays a critical role in energy expenditure and glucose homeostasis regulation [9, 10]. Skeletal muscle generates force through contraction and is necessary for movement, breathing, and posture maintenance [9]. Regarding its utility in plastic surgery, muscle transfer to remote areas can effectively treat open wounds post oncological resection, trauma, contractures, and severe contour defects [11, 12]. Furthermore, functional muscle can be transferred to create functional units in other areas of the body [13]. Other uses include tissue coverage of complex wounds decreasing the risk of bacterial infection while providing an adequate vascular bed for later skin grafting [14, 15], coverage following bone and cartilage debridement to

*A.C. Panayi and S.M. Lince contributed equally to this work.

A. C. Panayi, MD · I. Sinha, MD (✉)
Division of Plastic Surgery, Department of Surgery, Brigham & Women's Hospital, Harvard Medical School, Boston, MA, USA
e-mail: isinha@bwh.harvard.edu

S. M. Lince, BS
Department of Surgery, Penn State Hershey Medical Center, Hershey, PA, USA

G. Giatsidis (ed.), *Gene Therapy in Reconstructive and Regenerative Surgery*, https://doi.org/10.1007/978-3-319-78957-6_4

manage chronic osteomyelitis [16], coverage of difficult radiation wounds [17], and prophylactic wound coverage after early debridement and antibiotic treatment to salvage exposed prostheses [18]. Given the vast array of potential therapies involved with muscle flaps, augmentation of its mass and function could greatly improve clinical care, and GT may play a vital role in facilitating such advances.

4.1.1 Skeletal Muscle: Anatomy and Function

4.1.1.1 Skeletal Muscle Organization

The intricate structure of skeletal muscle is essential for optimal contractile function. Skeletal muscle is composed of individual muscle cells (myocytes/myofibers) gathered into bundles which contract as larger coordinated units held together by the endomysium, the basement membrane which surrounds individual myofibers, the perimysium which bundles multiple myofibers, and the epimysium which surrounds a complete muscle (Fig. 4.1). Fibers are surrounded by networks of nerves and blood vessels and an extracellular matrix (ECM) [9, 20]. Muscle stem cells, known as satellite cells (SCs), were discovered in 1961 between the plasma membrane (sarcolemma) of the myofiber and the basal membrane [21]. SCs are normally quiescent but when activated enter the cell cycle and proliferate and differentiate into myoblasts (partially differentiated intermediates between SCs and mature myocytes), which fuse to form multinucleated fibers with central nucleation; once the nuclei move to a subsarcolemmal position, they are referred to as myofibers [9, 22, 23].

While there is great overlap between embryonic myogenesis and the muscle healing process of SMR, there are discrepancies which may be amenable to GT intervention [19]. SCs have a unique embryonic origin—limb and trunk muscle SCs arise from the dermomyotome, whereas craniofacial SCs arise from the head mesoderm [24]. The number of SCs reaches a neonatal peak accounting for approximately a third of the total myofiber nuclei, a proportion which decreases to 2–7% of the total myofiber nuclei in adulthood [25, 26]. Neonatal SCs are more easily activated and more proliferative than adult SCs. Further understanding of the mechanisms underlying this phenomenon is a potential target for future GT intervention [27].

4.1.1.2 Skeletal Muscle Healing

Skeletal muscle injuries are common, nationally costly, and heterogeneous with respect to location and extent [28]. Skeletal muscle can better regenerate to native state than most other mature human tissues [20–22]. Yet when a large volume of muscle is lost due to skeletal muscle trauma, for example, volumetric muscle loss (VML) if the loss is ≥20% of the original mass of a given muscle, regeneration tends to be impaired with deposition of fibrous scar tissue and permanent loss of muscle volume [21, 29–31]. Given that the current available treatment options for VML are limited to scar tissue debridement or muscle transposition/transplantation and are associated with unfavorable outcomes, the RR surgeon has a pressing need to consider not only GT, TE, or both in synergy to advance traditional PRS options but also to improve VML management.

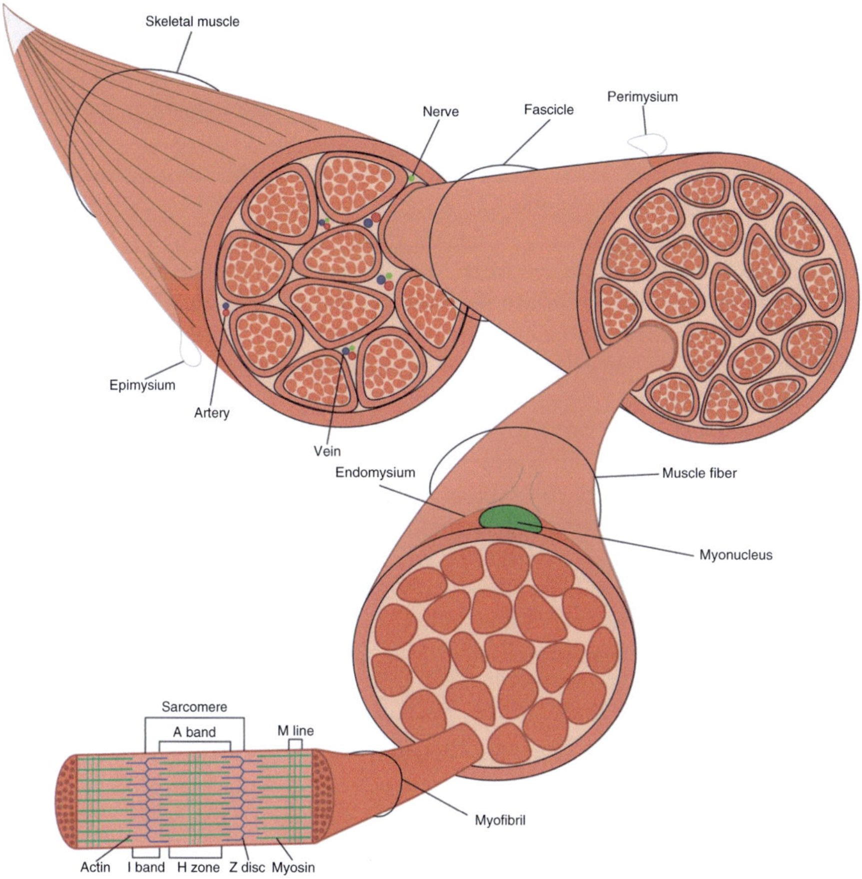

Fig. 4.1 Skeletal muscle organization and associated structures [19]

Skeletal muscle healing consists of three overlapping phases: (1) destruction, (2) repair, and (3) remodeling [9, 20, 22]. The destruction phase starts within minutes of injury and lasts up to 2 weeks. It is characterized by hematoma formation, muscle fiber necrosis, and increased local microvascular permeability with edema. This in turn causes macrophage and leukocyte influx, with the secretion of various growth factors and cytokines. This active signaling amplifies an intense local inflammatory response and sets up the microenvironment for the repair phase of healing [32]. The repair phase begins during the first week post-injury and lasts a variable length of time, during which necrotic debris is phagocytosed, new blood vessels are formed, and SCs regenerate mature myofibers [32, 33]. Quiescent Pax7-positive SCs migrate to the injury site and upregulate the myogenic regulatory factors (MRFs) MyoD and Myf5 to become proliferative—at which point they become referred to as myoblasts [34, 35]. Further myoblast differentiation steps include downregulation of Pax7 [36] and upregulation of the MRFs Mrf4 and myogenin [34]. After undergoing several rounds of amplification, myoblasts exit the cell cycle and either fuse to each other forming multinucleated myofibers (hyperplasia) or to the remaining myofibers to form terminally differentiated myofibers (hypertrophy) [22, 37]. SCs exhibit the crucial ability to self-renew and replenish the stem cell pool rather than all becoming activated [34, 38]. Both symmetric and asymmetric divisions have been proposed as mechanisms for SC self-renewal. Asymmetric SC division in skeletal muscles has been noted, whereby one daughter cell differentiates, while the other remains quiescent to continue self-renewal in an iterative process. Approximately 10% of the SCs never expressed MRF Myf5—suggesting that this population did not asymmetrically divide [38]. In observed symmetric division, all daughter SCs activate and proliferate, with a minority maintaining Pax7 expression while downregulating MyoD to contribute to the non-differentiating self-renewing portion [34]. The remodeling phase of muscle healing begins during the second week after muscle injury and lasts for a variable time during which regenerated myofibers mature and regain ability to contract; however, connective tissue reorganization can occur causing scar formation with contracture, inhibiting SMR [20, 32].

Thus, the destruction phase, immune/inflammatory response, and the remodeling phase, fibrosis/scarring, harbor many potential GT targets, but further understanding of the factors that influence SMR is necessary before discrimination between alternative options for research study design is possible—most currently suggested targets act during the repair phase.

4.1.2 Potential Gene Therapy Targets to Aid Skeletal Muscle Regeneration

4.1.2.1 Current Principle Gene Therapy Targets

Muscle SCs are the SMR precursor stem cells that have been best characterized in terms of regulatory mechanisms potentially amenable to GT approaches to date. The earliest known marker of SC activation is phosphorylated p38, followed by MyoD expression upregulation (Fig. 4.2) [40, 41]. MyoD is a challenging target for GT strategies, as it plays multiple context-specific roles in muscle differentiation and has been reported to inhibit SC proliferation in a differentiation-independent manner [42]. MyoD does not promote differentiation, but directly regulates Cdc6 expression during the G1 phase of the cell cycle rendering chromatin accessible for DNA replication [41]. MyoD can inhibit myoblast cell cycle entry by inducing p21 and upregulating apoptotic genes [43, 44]. Pax3, which is mainly expressed in the trunk and forelimb but not the hind limb muscles, and Pax7, which is expressed in SCs of all muscles, are SC markers, with Pax7 having the ability to prevent SCs from undergoing apoptosis [45, 46]. Pax7 expression levels within SCs are heterogeneous, with Pax7 "high" expression cells taking longer to enter the cell cycle than the roughly 3% Pax7 "low" SCs [47, 48]. Pax3 and Pax7 strongly impact Myf5 and MyoD but the mechanism through which they activate MyoD following muscular injury is yet to be elucidated [24]. One leading suggested mechanism is that Pax7 regulates inhibitor of DNA-binding (ID) proteins in quiescent SCs [49], and these

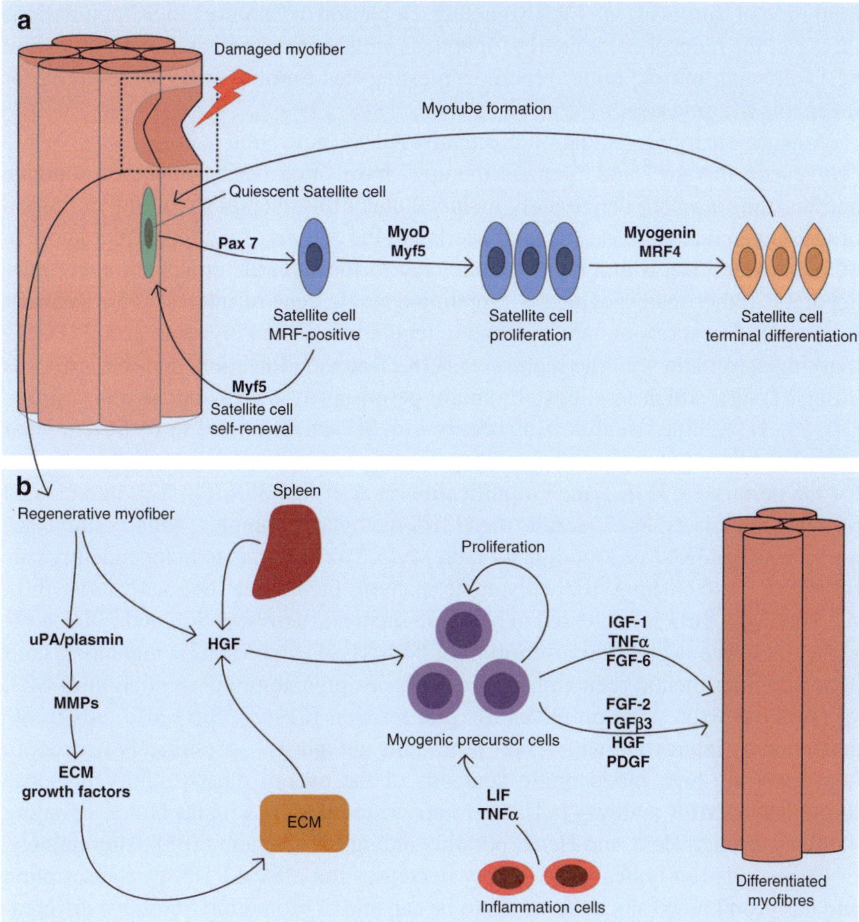

Fig. 4.2 (**a**) Satellite cell specification, activation, and differentiation during muscle development and regeneration. (**b**) The role of several secreted factors and extracellular matrix in muscle regeneration [39]

proteins are believed to be negative regulators of MyoD [50]. Other factors also seem to play significant independent roles in regulating SC quiescence versus activation. Membrane invaginations of quiescent SCs called caveolae express both caveolin-I, which regulates caveolae formation, and sphingomyelin, a plasma membrane-embedded lipid that facilitates cytoplasmic signaling by concentrating signaling molecules in the caveolae [51]. Calcitonin receptors are specifically present in quiescent SCs, suggesting that they could be involved in the process, but the mechanism has not yet adequately been elucidated [51]. Furthermore, the transmembrane protein Megf10 influences SC activation, with Megf10 gene silencing leading to differentiation and decreased proliferation, while overexpression enhances proliferation of SCs [52]. Another interesting potential GT target is inhibition of Sprouty1, via ERK signaling. Depletion of Sprouty1 in SC populations increases the number committed to apoptosis while concurrently reducing quiescent SCs following muscle injury repair, suggesting that Sprouty1 may be required for reversible SC quiescence [53].

A major emerging regulation mechanism of SC activation is epigenetics. While "epigenetic therapy" and "genetic therapy" have often been considered separate entities, they are both deliberately included under this chapter's working definition of GT. Given that there is extensive overlap of the actions of the major regulators in SC activation [54], it may be better strategy to focus on the epigenetic regulatory apparatus rather than specifically targeting a single gene of interest [55]. Relevant histone modifications include epigenetic regulation of the Polycomb group (PCG), marking chromatin with the repressive H3K27me3 modification, and the Trithorax group (TrxG), which modifies chromatin permissively increasing its transcription [56–58]. H3K27me3 modification occurs with SC activation and shifts SCs in favor of proliferation; genes promoting cell cycle progression are also reportedly enriched for the permissive H3K27me3 modification in activated SCs [58, 59]. In activated SCs and myoblasts, Pax7 recruits the H3K4 methylator complex, which establishes a permissive H3K27me3 modification on Myf5. Pax7 appears to independently participate in establishing the epigenetic pattern facilitating SC activation [60]. Alternatively, Sirt1 is expressed in both quiescent and activated SCs and can deacetylate and hence decrease transcription of both MyoD and MEF2D regulating their actions. Experimental reduction of Sirt1 induces premature differentiation of SCs, whereas overexpression promotes SC proliferation [61–63]. Sirt1 also aids in SC activation by interacting with ATG7 promoting autophagy—presumably in order to help meet the high bioenergetic demands of the overall muscle injury response through the AMPK pathway [64]. Sirt1 appears to interfere with the Notch signaling pathway through Hes1 and Hey2, possibly aiding SC activation [65]. Alternatively, DNA methylation typically reversibly decreases the affected DNA's transcription and has been known since the 1970s to be capable of promoting fibroblast differentiation to a myogenic lineage [66]. Hypermethylation of the DNA for the Notch1 promotor mediated by TNF-α and NF-KB leads to a decrease in SC proliferation and self-renewal. Treatment of isolated SCs with the DNA methyltransferase inhibitor sulforaphane downregulates the myostatin pathway, illustrating a potential opportunity to promote SC proliferation and differentiation, tangential to direct TGF-β family GT targeting [67].

Noncoding RNA makes up the majority of eukaryotic transcription products, and some of these products have been formally linked to skeletal muscle function and disease [68]. The myomiR family of microRNA is involved in muscle differentiation regulation and is capable of transdifferentiating fibroblasts to a muscular lineage [69]. Family member miR-1 expression is upregulated during muscular regeneration and directly represses Pax7 expression to promote SC proliferation and differentiation [70]. MicroRNAs outside the myomiR family also thought to be involved in SC activation include MiR-181, which is upregulated during muscular regeneration and targets negative regulation of MyoD HoxA11 promoting SC proliferation [71]. MiR-489, which represses Dek to inhibit cell cycle entry, is enriched in quiescent SC and is downregulated in activated SCs [72]. MiR-206, which represses expression of Pax7, Notch3, IGFBPS, and Hmgj3 to promote SC differentiation, is upregulated in activated SCs and myoblasts as well as in Duchenne muscles perhaps due to increased levels of SC activation [73–76]. Human SCs undergo global downregulation of microRNA expression when transitioning from quiescent to activated, most obvious in the cases of miR-1066, 25, 29c, 210c [77]. Unfortunately the majority of available studies on "long" noncoding RNA (>200 nucleotides [68]) have been conducted using the C2C12 myoblast cell line, but most, including Malat1, linc-MD1, SRA, Neat1, and YAM, appear to be upregulated and required for normal differentiation [78–82].

Muscle injury, stretching, overuse, and many degenerative muscle diseases induce infiltration of neutrophils and macrophages into skeletal muscle tissue [83]. Macrophage recruitment has been suppressed in vivo through injection of antibodies that inhibit binding of M-CSF to its receptor leading to incomplete SMR [84]. Furthermore, prevention of monocyte recruitment to the site of injury completely inhibits SMR, possibly due to decreased secretion of local factors as well as lack of direct macrophage-SC cell-cell interaction that protects SCs from apoptosis [84]. Consequently, the innate immune response plays a vital role in the response of skeletal muscle to injury [85] largely because cytokines, growth factors, and chemokines (specifically IL-6 and IFNγ) promote inflammation, attracting not only more neutrophils and macrophages but also platelets and B and T cells, and inducing fibroblast signaling, ultimately promoting SC activation and proliferation [86]. Authors have reported that M1/pro-inflammatory macrophages induce myogenic precursor cell proliferation, while M2/anti-inflammatory macrophages induce differentiation and fusion of these cells, underscoring the importance of sequential timing (M1 followed by M2) and thereby highlighting the potential superiority of GT strategies versus simple local factor treatment [87]. Autocrine factors from infiltrating inflammatory cells also influence SCs, which also respond to autocrine factors from other neighboring cells such as neurons [37]. While most regulatory factors involved in local inflammation are potential SMR GT targets, the specifics are not sufficiently understood to contemplate GT approaches beyond the principle cytokines and M1 and M2 differentiation signals.

4.1.2.2 Growth Factors

GT is uniquely well positioned to approach myogenesis or SMR-related problems, because in addition to the vast amount of available regulatory factors, the sequence and temporal release into the microenvironment also affects muscle generation (Table 4.1) [128]. As far as regulatory factor categories go, net stimulatory members include IGF1/IGF2, HGF, VEGF, FGF 2/FGF6, PDGF AA, and PDGF BB, and net inhibitory members include myostatin, TGF-α and TGF-β1, and BMPs (all members of TGF-β superfamily) [118].

The TGF-β superfamily comprises many secreted factors grouped into subfamilies according to sequence homology. Members influence nearly all aspects of cellular behavior and in general negatively regulate myogenesis by repressing MyoD and myogenin expression [88]. Myostatin is a TGF-β family member expressed in

Table 4.1 Key regulatory factors of interest for future investigation

Target	Systemic/ local	Major function	Unique features	Sources
Myostatin TGF-β family member	Systemic circulating	Inhibit muscle growth and differentiation; maintain SC quiescence	Concentrations increase with aging	[88–92]
Insulin-like growth factors Multiple family members	Systemic circulating	Maintain muscle mass; stimulate SC activation and SMR	Upstream of IGF-PI3K/Akt-mTOR-SGK axis	[93–97]
Fibroblast growth factors Multiple family members	Systemic circulating	Promote SC and myoblast cell cycle entry and proliferation likely through Ras signaling	Upregulation upon skeletal muscle injury correlates with SC activation	[98–103]
Vascular endothelial growth factors Multiple family members	Local	Promote angiogenesis and revascularization	Local therapy and GT have improved muscle flap hemodynamic properties	[104–113]
Hepatocyte growth factor	Local ECM-associated	Initiation step of SC activation; sequestered by ECM, released on muscle injury	Synthesized and secreted by many local cells including SCs	[114–116]
Heparin sulfate proteoglycans Multiple family members	Local ECM-associated	Aid SC activation through various signaling pathways	Present on both ECM and SC surface. Upregulated in SCs on muscle injury	[160–162, 117]
Matrix metalloproteinases Multiple family members	Local ECM-associated	ECM remodeling and SMR factor release on muscle injury; aid muscle precursor cell migration	MMP2 mainly via SCs-myofibers; MMP9 mainly via immune cells	[118–122]
Nitric oxide synthases Multiple family members	Local	Produce NO and promote HGF release from ECM; decrease ROS burden on muscle injury	iNOS isoforms upregulated during skeletal muscle injury	[123–127]

healthy muscle tissue that inhibits muscular growth and differentiation [89] and maintains SC quiescence via mechanisms such as MyoD repression through p21-cip induction as well as Pax7 promotion through ERK signaling [90, 91]. Aging increases TGF-β levels preventing SCs from entering the cell cycle due to downstream cell cycle inhibitors such as p15, p16, p21, and p27; in fact, treatment with local injection of anti-TNF-β family member antibodies allows SCs to reenter the cell cycle in greater numbers than normal in aged mice following muscular injury, establishing multiple points along this pathway as potential GT targets [92].

Insulin-like growth factors (IGFs) are circulating hormones upstream of the IGF-PI3K/Akt-mTOR-SGK axis, critical for development and regeneration of nearly every organ and vital for skeletal muscle mass maintenance [93]. Unfortunately the mechanism of IGF-mediated SC activation is not yet fully elucidated but has been suggested to involve upregulation of Myf5 upon injury either secondary to calcium influx (calcineurin and calmodulin kinase triggered through calcium binding to calmodulin activate Myf5) or through the PI3K/Akt and ERK signaling pathways [94]. Muscle damage induces alternative splicing of IGF, including MGF, local concentrations of which are specifically correlated with SC activation, elevating SOD and decreasing reactive oxygen species in the local skeletal muscle healing environment [95, 96]. The IGF isoform IGF-IEa which is expressed at a later time point in the SMR sequence better correlates with myoblast proliferation and differentiation [97].

Fibroblast growth factors (FGFs) are circulating heparin-binding regulators which bind to FGF-receptors 1–4, all of which are expressed in SCs [98, 99]. FGF receptor 1 is upregulated shortly after muscle injury correlating with SC activation [99]. Overexpression facilitates in vitro proliferation of myoblasts while simultaneously inhibiting their differentiation [100]. Of the FGF members, FGF6 is the one most upregulated during muscular regeneration. Homozygous FGF6 knockout mice have been observed to have lower numbers of activated SCs and a constant size of quiescent SCs [101]. Of interest to SMR GT, exogenous FGF1, 2, 4, 6, or 9 promote SC proliferation in vitro, thus hinting at a potential target for GT "rescue" or enhancement of SC proliferation in adverse scenarios [99]. FGF mechanisms of SC activation appear to work through the Ras signaling pathway as constant activation of Ras is known to bypass FGF promoting myoblast proliferation and by facilitating G1-to-S phase transition in myoblasts through the MKK-ERK signaling pathway [102, 103].

Nitric oxide (NO) is a freely diffusible small messenger molecule involved in many cellular functions. Dying myofibers release NO to potentiate hepatocyte growth factor (HGF)-based SC activation following muscle injury [123]. NO is synthesized in skeletal muscles by nitric oxide synthase (NOS), and iNOS mRNA levels are significantly increased in both damaged muscle fibers within 6 h post-injury and in the infiltrating macrophages, increasing measured NO levels local to the muscle injury site [124]. Homozygous iNOS knockout mice do possess SCs, but these fail to proliferate and differentiate following injury [125]. In the early stages of muscular regeneration, NO promotes macrophage reactive oxygen species-independent lysis of damaged fibers, and this reactive oxygen species independence prevents further local oxidative damage [126] and stimulates the release of HGF from the ECM, as well as secretion of other SC activating factors and cytokines from various immune cells [127]. In the later stages of muscular regeneration, NO protects the healing environment from reactive oxygen species produced from prolonged inflammation [124], providing yet another temporal example of the potential superiority of GT approaches to local regulatory factor control.

4.1.2.3 A Signaling Pathway Point of View

Notch signaling is involved in cell fate determination, yet the overlapping functions in this signaling pathway make predicting the outcome of GT manipulation challenging. Notch signaling is needed embryonically for myogenic cell fate commitment and maintenance of SC numbers, with Notch ligand delta-1 homozygous knockout mice displaying muscular hypertrophy secondary to premature SC differentiation [129]. Muscle-specific overexpression of Notch family member NICD increases Pax7-positive SC numbers but maintains them in an undifferentiated state [130]. On the other hand, depletion of Notch family member RBP-J decreases myogenic SC numbers due to increased embryonic differentiation [131]. Notch signaling thus appears essential for maintaining SC quiescence, and endogenous Notch signaling is greatly decreased in SCs of regenerating muscles within 20 h of injury [132]. A second wave of Notch signaling 4–5 days after injury appears to promote SC proliferation [133], and Notch3 homozygous knockout mice exhibit abnormally increased numbers of SCs [134], while Notch activation promotes proliferation of cultured SCs in vitro [135]. While the temporal aspect may be particularly relevant to GT approaches, the reader must first begin with the landmark sources to identify the members of the Notch signaling pathway most suitable as potential targets.

Wnt proteins are soluble molecules regulating multiple cellular processes that are upregulated following muscle damage and, like the Notch signaling pathway, are challenging potential GT targets due to complex interplays with pro- and anti-SMR regulatory pathways [135]. Exogenous Wnt 1, 3a, and 5a promote SC proliferation, while Wnt 4 and 6 inhibit SC proliferation [136], and β-catenin appears to promote SC self-renewal while simultaneously inhibiting their differentiation [137]. Other Wnt functions seem to mitigate SC proliferation: injection of Wnt inhibitors Dkk1 and sFRP3 into aged mice reduces the normally increased propensity toward fibrogenic differentiation of SCs in aged versus younger mice, presumably secondary to increased serum Wnt concentration with age [138]. Noncanonical Wnt pathway signaling also appears to influence SCs, with Wnt7a increasing the SC numbers and Wnt7a depletion markedly decreasing SC numbers through a positive feedback loop with fibronectin [117, 139].

4.1.3 Skeletal Muscle GT in Plastic and Reconstructive Surgery

There exist several reviews on GT for plastic and reconstructive surgery [6–8], yet few of these studies are concerned with the survival and outcomes of muscular flaps.

By far the most common GT target in muscle flap research has been vascular endothelial growth factor (VEGF). This focus may be due to the significance of VEGF in angiogenesis and consequently vascular delay, which has been used for over 500 years to increase the amount of tissue that can survive following surgical transfer in ischemic conditions [104]. VEGF was a key early target in GT, with its first major application in muscular tissue being a direct DNA injection that aids revascularization of ischemic rabbit limbs [105]. VEGF was subsequently used in the treatment of ischemic limbs in humans [106, 107]. In 1999 recombinant human VEGF was infused in a rat gracilis muscle flap exposed to induced ischemia, and improvement in the flap survival rate versus the control groups was noted, with less distortion in the histological architecture [108]. A decade later, rat cremaster muscle flaps were isolated on their neurovascular pedicles, and VEGF was infused before clamping the pedicles off for either 4 or 6 h of induced ischemia. While the flaps that sustained critical ischemia (6 h) had no significant microcirculatory flow in either the treatment or control groups, the flaps that sustained only subcritical ischemia (4 h) exhibited significantly better red blood cell velocity in arterioles and better capillary flow in the VEGF-treated groups [109]. Treatment with recombinant VEGF provided the necessary evidence to attempt GT application. In 2001, following a 4 h induction of ischemia in a rat gracilis muscle flap, a VEGF cDNA encoding plasmid plus lipofectamine (a cationic liposome) was injected directly into the muscle flap. Higher proportions of total flap survival in the treatment group versus the lipofectamine-only control group, as well as higher capillary count 7 days post-reperfusion, demonstrated that VEGF gene transfer exerts a protective effect on ischemic skeletal muscle [110]. More recently, it was reported that an AAV vector encoding human VEGF165 was used to transduce a pectineus muscular flap in an in vivo bioreactor chamber, in an attempt to use GT to supplement tissue engineering. Observations included enhanced muscular tissue formation visualized through histological analysis with an increase in arteriole number compared to the control tissue transfected only with the reported B-galactosidase gene [111].

Several other factors have been investigated using muscle flap models, though none influencing the factors or signaling molecules previously discussed which direct SMR regulation—leaving many exciting potential GT work ahead. In 2002, the cremaster muscle flap of a rat was subjected to ischemia, followed by angiopoietin-1 gene injection into the external iliac artery through adenovirus transfection. After periodic tissue extraction of the cremaster tube flap—preserved in a surgically created subcutaneous pocket in the lower limb—on days 3, 7, and 14 to evaluate microcirculatory measurements, there were no functional differences between treatment or control groups noted on day 3. However, by day 7 the functional capillary density was significantly higher, while the microvascular permeability index was significantly lower in the treated groups versus control groups, and at day 14 both these metrics were even more strikingly different between treatment and control groups. For the first time, the authors concluded that GT with angiopoietin-1 improved skeletal muscle flap hemodynamics and demonstrated the feasibility of intravascular gene therapy using a virus vector in an attempt to enhance muscle flap hemodynamics [112]. The same group published a highly interesting study later on in the same year comparing the efficacy of VEGF versus angiopoietin-1 for the enhancement of rat cremaster muscle flap perfusion in a unique, controlled translational trial. The study design included a PBS-treated group, a GT control group with an adenovirus transfecting a control reporter GFP, and groups with adenovirus transferring VEGF alone, angiopoietin-1 alone, or combined VEGF and angiopoietin. On both days 7 and 14, all treatment groups showed significantly higher numbers of capillary densities compared with the PBS and reporter GFP control groups. At day 14, the VEGF alone GT treatment group proved to be superior to both the angiopoietin-1 alone GT treatment group and the VEGF plus angiopoietin-1 GT treatment groups, despite a semi-linear increase in the number of functional capillaries in all treatment groups. The VEGF GT alone treatment group appeared to also have a higher increase in permeability. Overall, the potential of GT with the factors independently or in combination was demonstrated. All treatment groups exhibited roughly equal efficacy in augmenting muscle perfusion with more stable vessels 1 week after GT, with all treatment groups showing increased perfusion of the muscle flap over a period of 14 days highlighting a long-lasting effect of the GT. The study, however, suggested a slight superiority for GT with VEGF alone versus VEGF plus angiopoietin-1 [113]. These examples illustrate the potential of GT techniques to improve current muscle flap survival.

4.1.4 GT as a Supplement to Tissue Engineering of Skeletal Muscle

Advancements in muscle flap surgery have positioned the reconstructive surgeon at the forefront of clinical tissue engineering and vascularized composite allotransplantation [12]. GT may be of the greatest utility as a supplement to aid a tissue engineering approach to manufacture functional skeletal muscle. Plastic and reconstructive literature has suggested a future involving prefabrication of grafts through tissue engineering, minimizing donor site morbidity [140, 141]. GT could potentially play a role in overcoming some of the limitations that tissue engineering has encountered thus far [7]. As with GT delivery methods, the current scope of muscular and related tissue engineering is sufficiently vast that entire texts are devoted to the subject [142], and exciting results on the manufacture of muscular flaps utilizing tissue engineering techniques are now being published [143–146]. GT and tissue engineering of skeletal muscle are rarely explored simultaneously, and hence a brief discussion of tissue engineering to emphasize classic fundamental problems GT techniques may be poised to address is imperative.

Interest in skeletal muscle tissue engineering increased with elucidation of the fact that myoblasts drive SMR; however, early clinical trials transplanting these partially differentiated muscle precursor cells provided lackluster results. Early myoblast injection-transplantation studies after in vitro expansion demonstrated feasibility of the concept, but unfortunately most myoblasts were lost within 1 week with low incorporation into the SC niche [147]. Subsequent studies demonstrated that injection-transplantation of the upstream SCs pre-activation greatly improved transplantation efficiency with far superior functional outcomes compared to myoblast control groups [148]. The local SC microenvironment or "niche" became appreciated as a unique combination of cellular, biophysical, and biochemical components, and the net influence of this confluence greatly impacted overall muscle regeneration, thus skeletal muscle tissue engineering strategies focused intensely on this niche.

4.1.4.1 The Satellite Cell Niche

While most tissue engineering strategies have been "cell centric," [29] the last major known actor to be discussed here once the principle cells, regulatory factors, and signaling pathways are appreciated is the ECM. The ECM exerts a profound influence on gene expression of many local cells by maintaining the SC niche and helping the local cells respond to alterations in the local microenvironment [149]; thus, the ECM must be considered and accounted for by any skeletal muscle GT strategy. Disruption of the SC niche is considered a pivotal event in the transition from SC quiescence to activation. Typically, in skeletal muscle injury, both disruption of the protective SC niche and direct signals from dying muscle fibers seem to play major roles in signaling SC activation. Mechanistically, intracellular SC calcium influx, hepatocyte growth factor (HGF) released from the disrupted ECM, and mobilization of normally α7/β1 integrin-anchored SCs from the collagen-laminin network are thought to play a role, thus all representing possible GT strategy targets [150, 151]. Yet selective muscle fiber damage using bupivacaine without basal lamina disruption also results in SC activation, suggesting that damaged fibers contribute signals in the activation of quiescent SCs during muscular injury independent of ECM/SC niche disruption [152]. SCs express integrin α7β1 basally (which links SC cytoskeleton with basement membrane laminin and is vital for the crucial role of transduction of mechanical forces into chemical signals), while SCs express M-cadherin apically (which attaches SCs to the myofiber); both attachment sites are essential for signal transduction between the SC in its niche and the two major flanking structures [118]. Some of the emerging mechanistic evidence involving SMR suggests that ECM-mediated functions involve both recruitment of stem/progenitor cells [153, 154] and modulation of macrophage phenotype [85, 155]. One study of interest with an informative discussion implicated many of these local factors simultaneously, showing that integrin α7β1 is vital for SC migration, that HGF plays a vital role in the guidance of SCs, and that post-division (but unrelated) SCs remain in contact with each other and co-migrate along the lengths of myofibers [156].

Several regulatory factors are best described in conjunction with a discussion on the SC niche, due to their close relationship to the ECM. HGF is a significant signaling molecule with potential implications for future GT applications, and a well-characterized example of one of many growth factors that the ECM "sequesters," in this case using HSPGs [114]. HGF is a mesenchymal heparin-binding glycoprotein which regulates cell division, motility, morphogenesis, and organ regeneration by binding the cMET receptor and activating a tyrosine kinase cascade [115]. SCs, myocytes, and other local cells all secrete HGF, which the ECM sequesters until muscle injury prompts its release; HGF has been shown to prominently activate SCs thus facilitating both paracrine and autocrine SC activation [150]. HGF-bound cMET has been suggested to be the initial SC activation step and marker, converting Go SCs to "alert" via HGF/cMET/mTORC1 signaling [116]. Two other entities of interest to potential future GT-SMR strategies that intimately interact with the ECM are heparan sulfate proteoglycans (HSPGs) and matrix metalloproteinases (MMPs). HSPG family members mainly vary in their extracellular domains which aid in tethering extracellular factors in order to "sequester" them and are found not only in the ECM but also on the surface of SCs and become upregulated in SCs upon activation [157]. SCs specifically express ECM-proteins syndecan 3 and 4 [157], transmembrane HSPGs which serve as low-affinity FGF receptors [158]. SC activation is accelerated but then arrested in the S phase in syndecan-3 homozygous knockout mice [159]. In contrast, SC activation requires syndecan-4, as homozygous knockout mice have a loss of MAPK signaling [159]. Syndecan can also facilitate noncanonical Wnt signaling by forming a co-receptor complex with Frizzled-7 [117], implying a complex regulatory signaling overlap. On the other hand, MMPs capable of degrading ECM proteins are released at sites of muscular injury by the damaged myofibers and SCs which release predominately MMP2 [119, 120] and by the infiltrating macrophages and leukocytes which release predominantly MMP9 [120]. NO increases MMP2 and MMP9 expression and activity levels, both cleaving collagens, laminins, and other ECM components to release HGF from sequestered ECM sites [121, 122]. MMPs are also involved in myoblast migration [118].

4.1.4.2 Future Prospects

Early skeletal muscle tissue engineering focused on myogenic cell proliferation through terminal differentiation into myotubules within a specialized biocompatible and biodegradable 3D support matrix or scaffold. Scaffolds provide a nurturing environment for the myoblast attachment, proliferation, alignment, and differentiation. Much of the success depended upon artificial replication or alteration of conditions found within the SC niche during periods of myogenesis or SMR.

Mechanical support is a major function of the ECM. The elastic stiffness of the ECM, often measured in terms of Young's modulus, strongly influences SC proliferation and differentiation [160], a mechanical principle incorporated into gel matrices and bioscaffolds. Consequently, the native SC niche is replicated alongside impregnated (time-released) regulatory factors and thus hopefully enabling muscle regeneration [161]. The importance of attention to successful recapitulation of the native SC niche was highlighted by experiments in which single myofiber transplantation (containing the rough niche) was more effective in restoring skeletal muscle function than myogenic stem cell transplantation not only in terms of number of myonuclei generated but also capacity for subsequent self-renewal [162].

Another major requirement for tissue engineering strategies is that the support material's porosity and permeability allow efficient vascularization for oxygen and other nutrient delivery, for waste removal, and for cell migration. Early gel matrices included Matrigel with collagen (ECM proteins along with undefined concentrations of growth factors to enhance myogenic differentiation, which facilitated generation of parallel arrays of contraction-capable myofibers with small diameters) [163] and fibrin within a gel matrix (allows space for ECM generation as well as diffusion of nutrients for in vitro cultures; it is of great interest to this chapter that thrombin incorporation allowed for superior diffusion enhancing cellular growth [164] and has since been successfully employed to enhance local SMR in rodent studies) [165]. Gel matrices usually lack the structural organization for adequate contractile force generation required by the muscular tissue. In contrast, scaffold supports made from biopolymers and synthetic polymers are more biocompatible than the listed matrices, as well as more easily handled and surgically attached. Early ECM-coated PLLA scaffolds enabled organization of functional muscle with myofibers into parallel orientation [166], while PGA mesh allowed highly vascularized muscle-like tissue to develop in rat models after 6 weeks [167]. DegraPol, an electrospun degradable polyesterurethane membrane, was originally used to engineer skeletal muscle with multinucleated myotubules expressing myogenic markers, which allowed C2C12 myoblast adhesion and myotube alignment along with scaffold fibers and degraded within 180 days [168]. Biologic scaffolds are now a major strategy being pursued in the engineering of skeletal muscle with a translational focus on VML [29]. Surgical placement of mammalian ECM has been reported to promote functional SMR following experimentally induced skeletal muscle injury in animal models [169–171]. There have also been reports of successful clinical use of ECM scaffolds in human patients after extremity trauma with VML [29, 153].

Several classic problems associated with bioscaffolds but also relevant to newer approaches such as additive manufacturing, or 3D bioprinting, are likely to be addressable with GT based on specific study findings. A major problem with tissue engineering of skeletal muscle is how to properly vascularize the interior of the 3D units during muscle tissue growth, in order to maintain essential nutrition supply and waste disposal that occurs by simple diffusion at the peripheries of the unit. To address this specific challenge, Langer et al. seeded a porous biodegradable PLLA and PGA scaffold with embryonic fibroblasts and endothelial cells along with myoblasts. The embryonic fibroblasts secreted VEGF—which induced the endothelial cells to produce functional blood vessels [172–174]. Combined with the results of GT studies of VEGF on muscular flaps, this could be an obvious opportunity where GT may facilitate tissue engineering of thicker skeletal muscle either in vitro pre-transplantation or in vivo. During an extensive skeletal muscular injury, the muscular blood vessels are disrupted, with vascular endothelial cells releasing growth factors such as VEGF, PDGF, and HGF, all of which promote SC activation and proliferation while supporting vessel hemostasis, revascularization, and angiogenesis [47]. Most SCs in humans (68%) and mice (82%) are located within 5 μm from neighboring capillaries or vascular endothelial cells, and there is a correlation between the number of capillaries per muscle fiber and the number of SCs, suggesting an intimate SC-endovascular cell relationship during myogenesis [175]. All are potential future targets for GT augmentation of skeletal muscle tissue engineering. In addition to vascular supply, scaffold-based tissue-engineered skeletal muscle suffers from another major limitation which GT may ameliorate: although perhaps superior to matrix gels, forces generated by fabricated constructs still tend to be 2–10% lower than adult mammalian skeletal muscle [176]. Yet studies have shown that the force generated by engineered muscle can be significantly improved by adding local IGF-1, primarily due to an increase in myofiber size (~40% increase), contractile protein synthesis, and improved sarcomeric organization [177].

While bioscaffolds have been impregnated with growth factors to facilitate the engineering of thicker tissue products, less attention has been given to the strategy of utilizing GT for stable production of not only growth factors but also cytokines and other paracrine factors. GT in this manner would be more technically challenging than the impregnation of tissue engineering support devices, yet the timing and dosing of various regulatory factors could in principle be far more rigorously controlled in order to optimize outcomes. In addition, a wider variety of factors could be considered with less constraints on the biomaterials since these materials would not have to be selected on the basis of their ability to be impregnated with the factor(s) of interest. In addition, nuanced tissue engineering applications in plastic and reconstructive surgery—such as for accurate craniofacial reconstruction—demands not only production of complex, multicell-type tissues with high levels of organization and polarity but also tissue interfaces for vital structures such as muscular attachments. Manufacture of such interfaces requires multi-tissue coordination, and identification and controlled delivery of complex signals in both time and space, making GT a likely supplemental aid to regulating some of the finer points of future, chimeric tissue engineering [8]. However, the actualization of goals such as these requires a much broader pallet of muscle-based GT options, supported through studies investigating GT with many of the regulatory factors involved in physiologic muscle generation and regeneration.

4.2 Summary

Plastic and reconstructive surgery has a record of utilizing skeletal muscle in an increasing variety of roles and a long-standing interest in minimizing donor site need for autologous mobilization procedures. Consequently, there is inherent interest in GT as it offers the potential of constructing functional skeletal muscle without the need of a donor site in situations such as VML where skeletal muscle required. GT has evolved into a set of techniques with immense potential to alter almost all regulatory factors, such as silencing expression of a factor at a specific stage of muscle generation or facilitating expression of an influential factor at specific time and level—even in cell types that are typically local bystanders.

Yet to date, only a small number of factors have been empirically tested in controlled GT studies on skeletal muscle—the academic surgical community has not fully investigated what we already have, and the currently available literature on the subject reflects this. Most GT applications explored in plastic reconstructive reviews focus on a superficial comparison of currently available GT techniques, and describes successful completed studies in great detail, without reviewing the far more numerous regulatory factors recognized in the process of SMR for future research. On the other hand, most skeletal muscle tissue engineering reviews do not consider GT prominently, if at all. Furthermore, nearly all studies and reviews on GT use in skeletal muscle at the translational or clinical level focus solely on remediation of single-gene inherited muscular diseases, thus neglecting the overwhelming majority of conceivable medical or surgical applications of GT in skeletal muscle generation.

Plastic and reconstructive surgeons should attempt to bring together the vast wealth of knowledge on muscular generation in order to be better able to select from the large range of possible GT targets, in order to either directly apply this to surgical problems or to confront classic problems of functional skeletal muscle regeneration. This review has attempted to provide specific details on each issue raised, while simultaneously establishing a panoramic view in order to provide the reader with an extensive perspective, providing a good starting point from which to start follow-up inquiries to generate informed study hypotheses and concepts of GT tailored to specific reconstructive problems.

References

1. Wolff JA, et al. Science. 1990;247:1465–8.
2. Tabebordbar M, et al. Science. 2015;5177:1–9.
3. Nelson CE, et al. Science. 2015;5143:1–8.
4. Duan D. Muscle gene therapy. Berlin: Springer; 2014. (ISBN 1489985247).
5. Duan D. Muscle gene therapy, methods and protocols. New York: Humana Press; 2011. (ISBN 1617379816).
6. Tepper OM, Mehrara BJ. Plast Reconstr Surg. 2002;109:716–34.
7. Roman S, Lindeman R, O'Toole G, Poole MD. Curr Gene Ther. 2005;5:81–99.
8. Scheller EL, Villa-Diaz LG, Krebsbach PH. J Craniofac Surg. 2012;23:333–7.
9. Huard J, Li Y, Fu FH. J Bone Joint Surg Am. 2002;84-A:822–32.
10. Stump CS, Henriksen EJ, Wei Y, Sowers JR. Ann Med. 2006;38:389–402.
11. Mathes SJ. Clinical applications for muscle and musculocutaneous flaps. 2nd ed.; 1981. ISBN:801631645.
12. Thorne CH, Chung KC, Gosain A, et al. Grabb and Smith's plastic surgery. Philadelphia: LWW; 2013. (ISBN 1451109555).
13. Stevanovic M, Sharpe F. Atlas Hand Clin. 2002;7:163–80.
14. Calderon W, Chang N, Mathes SJ. Plast Reconstr Surg. 1986;77:785–92.
15. Gosain A, Chang N, Mathes S, Hunt TK, Vasconez L. Plast Reconstr Surg. 1990;86(6):1152–62. Available at: http://www.ncbi.nlm.nih.gov/pubmed/2243859
16. Mathes SJ, Alpert BS, Chang N. Plast Reconstr Surg. 1982;69:815–29.
17. Mathes SJ, Alexander J. Surg Oncol Clin N Am. 1996;5:809–24.
18. Greenberg B, LaRossa D, Lotke PA, Murphy JB, Noone RB. Plast Reconstr Surg. 1989;83:85–89,97–99.
19. Tajbakhsh S. J Intern Med. 2009;266:372–89.
20. Järvinen TAH, Järvinen TLN, Kääriäinen M, Kalimo H, Järvinen M. Am J Sports Med. 2005;33:745–64.
21. Mauro A. J Biophys Biochem Cytol. 1961;9:493–5.
22. Chargé SBP, Rudnicki MA. Physiol Rev. 2004;84:209–38.
23. Shi X, Garry D. Genes Dev. 2006;20:1692–708.
24. Fu X, Wang H, Hu P. Stem cell activation in skeletal muscle regeneration. Cell Mol Life Sci. 2015;72(9):1663–77.
25. Allbrook DB, Han MF, Hellmuth AE. Pathology. 1971;3:223–43.
26. Schmalbruch H, Hellhammer U. Anat Rec. 1976;185:279–87.
27. Schultz E. Dev Biol. 1996;175:84–94.
28. Jarvinen TAH, et al. Muscle injuries: optimising recovery. Best Pract Res Clin Rheumatol. 2007;21:317–31.
29. Sicari BM, et al. Sci Transl Med. 2014;6:234ra58.
30. Grogan BF, Hsu JR. J Am Acad Orthop Surg. 2011;19(Suppl 1):S35–7.
31. Corona BT, et al. Biomaterials. 2013;34:3324–35.
32. Ceafalan LC, Popescu BO, Hinescu ME. Cellular players in skeletal muscle regeneration. Biomed Res Int. 2014;2014:957014. https://doi.org/10.1155/2014/957014.
33. Zammit PS, Partridge TA, Yablonka-Reuveni Z. J Histochem Cytochem. 2006;54:1177–91.
34. Cornelison DD, Wold BJ. Dev Biol. 1997;191:270–83.
35. Beauchamp JR, et al. J Cell Biol. 2000;151:1221–33.

36. Zammit PS, et al. J Cell Biol. 2004;166:347–57.
37. Hawke TJ, Garry DJ. J Appl Physiol. 2001;91:534–51.
38. Kuang S, Kuroda K, Le Grand F, Rudnicki MA. Cell. 2007;129:999–1010.
39. Karalaki M, Fili S, Philippou A, Koutsilieris M. Muscle regeneration: cellular and molecular events. In Vivo. 2009;23:779–96.
40. Jones NC, et al. J Cell Biol. 2005;169:105–16.
41. Zhang K, Sha J, Harter ML. J Cell Biol. 2010;188:39–48.
42. Sorrentino V, Pepperkok R, Davis RL, Ansorge W, Philipson L. Nature. 1990;345:813–5.
43. Halevy O, et al. Science. 1995;267:1018–21.
44. Asakura A, et al. Proc Natl Acad Sci U S A. 2007;104:16552–7.
45. Relaix F, et al. J Cell Biol. 2006;172:91–102.
46. Zanou N, Gailly P. Skeletal muscle hypertrophy and regeneration: interplay between the myogenic regulatory factors (MRFs) and insulin-like growth factors (IGFs) pathways. Cell Mol Life Sci. 2013;70:4117–30.
47. Montarras D, L'Honoré A, Buckingham M. Lying low but ready for action: the quiescent muscle satellite cell. FEBS J. 2013;280:4036–50.
48. Pallafacchina G, et al. Stem Cell Res. 2010;4:77–91.
49. Kumar D, Shadrach JL, Wagers AJ, Lassar AB. Mol Biol Cell. 2009;20:3170–7.
50. Boldrin L, Zammit PS, Muntoni F, Morgan JE. Stem Cells. 2009;27:2478–87.
51. Gnocchi VF, White RB, Ono Y, Ellis JA, Zammit PS. PLoS One. 2009;4:e5205. https://doi.org/10.1371/journal.pone.0005205.
52. Holterman CE, Le Grand F, Kuang S, Seale P, Rudnicki MA. J Cell Biol. 2007;179:911–22.
53. Shea KL, et al. Cell Stem Cell. 2010;6:117–29.
54. Yao S, He Z, Chen C. Hum Gene Ther. 2015;26:463–71.
55. Frank O, et al. Blood. 2004;104:3543–9.
56. Lee H, Habas R, Abate-Shen C. Science. 2004;304:1675–8.
57. Hansen KH, et al. Nat Cell Biol. 2008;10:1291–300.
58. Barski A, et al. Cell. 2007;129:823–37.
59. Palacios D, et al. Cell Stem Cell. 2010;7:455–69.
60. McKinnell IW, et al. Nat Cell Biol. 2008;10:77–84.
61. Fulco M, et al. Dev Cell. 2008;14:661–73.
62. Fulco M, et al. Mol Cell. 2003;12:51–62.
63. Zhao X, Sternsdorf T, Bolger TA, Evans RM, Yao T-P. Mol Cell Biol. 2005;25:8456–64.
64. Tang AH, Rando TA. EMBO J. 2014;33:2782–97.
65. Guarani V, et al. Nature. 2011;473:234–8.
66. Taylor SM, Jones PA. Cell. 1979;17:771–9.
67. Fan H, et al. Epigenetics. 2012;7:1379–90.
68. Erriquez D, Perini G, Ferlini A. Non-coding RNAs in muscle dystrophies. Int J Mol Sci. 2013;14:19681–704.
69. Kirby TJ, McCarthy JJ. MicroRNAs in skeletal muscle biology and exercise adaptation. Free Radic Biol Med. 2013;64:95–105.
70. Chen JF, et al. J Cell Biol. 2010;190:867–79.
71. Naguibneva I, et al. Nat Cell Biol. 2006;8:278–84.
72. Cheung TH, et al. Nature. 2012;482:524–8.
73. Cacchiarelli D, et al. EMBO Rep. 2011;12:136–41.
74. van Rooij E, Liu N, Olson EN. MicroRNAs flex their muscles. Trends Genet. 2008;24:159–66.
75. Chen J-F, Callis TE, Wang D-Z. J Cell Sci. 2009;122:13–20.
76. Ge Y, Chen J. MicroRNAs in skeletal myogenesis. Cell Cycle. 2011;10:441–8.
77. Koning M, Werker PMN, van Luyn MJA, Krenning G, Harmsen MC. Differentiation. 2012;84:314–21.
78. Legnini I, Morlando M, Mangiavacchi A, Fatica A, Bozzoni I. Mol Cell. 2014;53:506–14.
79. Watts R, Johnsen VL, Shearer J, Hittel DS. Am J Physiol Cell Physiol. 2013;304:C995–1001.
80. Caretti G, et al. Dev Cell. 2006;11:547–60.
81. Lu L, et al. EMBO J. 2013;32:2575–88.

82. Twayana S, et al. Biochem Soc Trans. 2013;41:844–9.
83. Toumi H, F'guyer S, Best TM. The role of neutrophils in injury and repair following muscle stretch. J Anat. 2006;208:459–70.
84. Segawa M, et al. Exp Cell Res. 2008;314:3232–44.
85. Tidball JG, Villalta SA. Am J Physiol Regul Integr Comp Physiol. 2010;298:R1173–87.
86. Kharraz Y, Guerra J, Mann CJ, Serrano AL, Muñoz-Cánoves P. Macrophage plasticity and the role of inflammation in skeletal muscle repair. Mediat Inflamm. 2013;2013:491497. https://doi.org/10.1155/2013/491497.
87. Arnold L, et al. J Exp Med. 2007;204:1057–69.
88. Kollias HD, McDermott JC. J Appl Physiol. 2008;104:579–87.
89. Joulia-Ekaza D, Cabello G. The myostatin gene: physiology and pharmacological relevance. Curr Opin Pharmacol. 2007;7:310–5.
90. McCroskery S, Thomas M, Maxwell L, Sharma M, Kambadur R. J Cell Biol. 2003;162:1135–47.
91. McFarlane C, et al. Exp Cell Res. 2008;314:317–29.
92. Carlson ME, Hsu M, Conboy IM. Nature. 2008;454:528–32.
93. Sandri M, et al. Cell. 2004;117:399–412.
94. Perez-Ruiz A, Gnocchi VF, Zammit PS. Cell Signal. 2007;19:1671–80.
95. Dobrowolny G, et al. J Cell Biol. 2005;168:193–9.
96. Owino V, Yang SY, Goldspink G. FEBS Lett. 2001;505:259–63.
97. Hill M, Goldspink G. J Physiol. 2003;549:409–18.
98. Kästner S, Elias MC, Rivera AJ, Yablonka-Reuveni Z. J Histochem Cytochem. 2000;48:1079–96.
99. Sheehan SM, Allen RE. J Cell Physiol. 1999;181:499–506.
100. Scata KA, Bernard DW, Fox J, Swain JL. Exp Cell Res. 1999;250:10–21.
101. Floss T, Arnold HH, Braun T. Genes Dev. 1997;11:2040–51.
102. Fedorov YV, Scott Rosenthal R, Olwin BB. J Cell Biol. 2001;152:1301–5.
103. Jones NC, Fedorov YV, Rosenthal RS, Olwin BB. J Cell Physiol. 2001;186:104–15.
104. Ghali S, Butler PEM, Tepper OM, Gurtner GC. Plast Reconstr Surg. 2007;119:1735–44.
105. Takeshita S, et al. J Clin Invest. 1994;93:662–70.
106. Isner JM, et al. Lancet. 1996;348:370–4.
107. Baumgartner I, et al. Circulation. 1998;97:1114–23.
108. Kryger Z, et al. Ann Plast Surg. 1999;43:172–8.
109. Banbury I, Siemionow M, Porvasnik S, Petras S, Browne E. Plast Reconstr Surg. 2000;106:1–6.
110. Neumeister MW, Song YH, Mowlavi A, Suchy H, Mathur A. Microsurgery. 2001;21:58–62.
111. Moimas S, et al. J Tissue Eng. 2015;6:2041731415611717. https://doi.org/10.1177/2041731415611717.
112. Gurunluoglu R, Lubiatowski P, Goldman CK, Carnevale K, Siemionow M. Ann Plast Surg. 2002;48:401–9.
113. Lubiatowski P, et al. Plast Reconstr Surg. 2002;110:149–59.
114. Kresse H, Schnherr E. Proteoglycans of the extracellular matrix and growth control. J Cell Physiol. 2001;189:266–74.
115. Zarnegar R, Michalopoulos GK. The many faces of hepatocyte growth factor: from hepatopoiesis to hematopoiesis. J Cell Biol. 1995;129:1177–80.
116. Leshem Y, Spicer DB, Gal-Levi R, Halevy O. J Cell Physiol. 2000;184:101–9.
117. Bentzinger CF, et al. Cell Stem Cell. 2013;12:75–87.
118. Ten Broek RW, Grefte S, Von Den Hoff JW. Regulatory factors and cell populations involved in skeletal muscle regeneration. J Cell Physiol. 2010;224:7–16.
119. Fukushima K, et al. BMC Musculoskelet Disord. 2007;8:54.
120. Kherif S, et al. Dev Biol. 1999;205:158–70.
121. Yamada M, et al. Int J Biochem Cell Biol. 2008;40:2183–91.
122. Yamada M, et al. Muscle Nerve. 2006;34:313–9.
123. Wozniak AC, Anderson JE. Dev Dyn. 2007;236:240–50.

124. Stamler JS, Meissner G. Physiol Rev. 2001;81:209–37.
125. Rigamonti E, et al. J Immunol. 2013;190:1767–77.
126. Tidball JG. AJP Regul Integr Comp Physiol. 2004;288:R345–53.
127. Ganea E, Trifan M, Laslo AC, Putina G, Cristescu C. Biochem Soc Trans. 2007;35:689–91.
128. Hayashi S, et al. Histochem Cell Biol. 2004;122:427–34.
129. Schuster-Gossler K, Cordes R, Gossler A. Proc Natl Acad Sci U S A. 2007;104:537–42.
130. Mourikis P, Gopalakrishnan S, Sambasivan R, Tajbakhsh S. Development. 2012;139:4536–48.
131. Vasyutina E, et al. Proc Natl Acad Sci U S A. 2007;104:4443–8.
132. Bjornson CRR, et al. Stem Cells. 2012;30:232–42.
133. Mourikis P, Tajbakhsh S. BMC Dev Biol. 2014;14:2.
134. Delfini MC, Hirsinger E, Pourquié O, Duprez D. Development. 2000;127:5213–24.
135. Brack AS, Conboy IM, Conboy MJ, Shen J, Rando TA. Cell Stem Cell. 2008;2:50–9.
136. Yin H, Price F, Rudnicki MA. Physiol Rev. 2013;93:23–67.
137. Polesskaya A, Seale P, Rudnicki MA. Cell. 2003;113:841–52.
138. Brack AS, et al. Science. 2007;317:807–10.
139. Le Grand F, Jones AE, Seale V, Scimè A, Rudnicki MA. Cell Stem Cell. 2009;4:535–47.
140. Garfein ES, Orgill DP, Pribaz JJ. Clinical applications of tissue engineered constructs. Clin Plast Surg. 2003;30:485–98.
141. Pribaz IJ, Fine NA. Clin Plast Surg. 2001;28(2):261–72, vii. Available at: http://www.ncbi.nlm.nih.gov/pubmed/11400820
142. Laurencin CT, Khan Y. Regenerative engineering. Boca Raton: CRC Press; 2013. (ISBN 9781782423140).
143. Shandalov Y, et al. Proc Natl Acad Sci U S A. 2014;111:6010–5.
144. Shandalov Y, Egozi D, Freiman A, Rosenfeld D, Levenberg S. Methods. 2015;84:70–5. https://doi.org/10.1016/j.ymeth.2015.03.021.
145. Egozi D, et al. J Vis Exp. 2016;107:52984. https://doi.org/10.3791/52984.
146. Zhao Y, Feric NT, Thavandiran N, Nunes SS, Radisic M. The role of tissue engineering and biomaterials in cardiac regenerative medicine. Can J Cardiol. 2014;30:1307–22.
147. Otto A, Collins-Hooper H, Patel K. The origin, molecular regulation and therapeutic potential of myogenic stem cell populations. J Anat. 2009;215:477–97.
148. Sacco A, Doyonnas R, Kraft P, Vitorovic S, Blau HM. Nature. 2008;456:502–6.
149. Bissell MJ, Hall HG, Parry G. J Theor Biol. 1982;99:31–68.
150. Tatsumi R, Anderson JE, Nevoret CJ, Halevy O, Allen RE. Dev Biol. 1998;194:114–28.
151. Maina F, et al. Cell. 1996;87:531–42.
152. Bischoff R. Development. 1990;109:943–52.
153. Mase VJ, et al. Orthopedics. 2010;33:511.
154. Agrawal V, et al. Proc Natl Acad Sci U S A. 2010;107:3351–5.
155. Badylak SF, Valentin JE, Ravindra AK, McCabe GP, Stewart-Akers AM. Tissue Eng Part A. 2008;14:1835–42.
156. Siegel AL, Atchison K, Fisher KE, Davis GE, Cornelison DDW. Stem Cells. 2009;27:2527–38.
157. Cornelison DD, Filla MS, Stanley HM, Rapraeger AC, Olwin BB. Dev Biol. 2001;239:79–94.
158. Quinlan JG, et al. Muscle Nerve. 1995;18:201–6.
159. Cornelison DDW, et al. Genes Dev. 2004;18:2231–6.
160. Gilbert PM, et al. Science. 2010;329:1078–81.
161. Discher DE, Mooney DJ, Zandstra PW. Science. 2009;324:1673–7.
162. Collins CA, et al. Cell. 2005;122:289–301.
163. Powell CA, Smiley BL, Mills J, Vandenburgh H. Am J Physiol Cell Physiol. 2002;283:C1557–65.
164. Koning M, Harmsen MC, van Luyn MJA, Werker PMN. J Tissue Eng Regen Med. 2009;3:407–15.
165. Beier JP, et al. Plast Reconstr Surg. 2006;118(5):1113–21. https://doi.org/10.1097/01.prs.0000221007.97115.1d.
166. Cronin EM, et al. J Biomed Mater Res A. 2004;69:373–81.

167. Saxena AK, Willital GH, Vacanti JP. Biomed Mater Eng. 2001;11:275–81.
168. Riboldi SA, Sampaolesi M, Neuenschwander P, Cossu G, Mantero S. Biomaterials. 2005;26:4606–15.
169. Sicari BM, et al. Tissue Eng Part A. 2012;18:1941–8.
170. Corona BT, et al. Tissue Eng Part A. 2012;18:1213–28.
171. Turner NJ, et al. Tissue Eng Part A. 2010;16:3309–17.
172. Levenberg S, et al. Differentiation of human embryonic stem cells on three-dimensional polymer scaffolds. Proc Natl Acad Sci U S A. 2003;100:12741–6.
173. Levenberg S, et al. Nat Biotechnol. 2005;23:879–84.
174. Huang Y-C, Dennis RG, Larkin L, Baar K. J Appl Physiol. 2005;98:706–13.
175. Christov C, et al. Mol Biol Cell. 2007;18:1397–409.
176. Moon du G, Christ G, Stitzel JD, Atala A, Yoo JJ. Tissue Eng Part A. 2008;14:473–82.
177. Vandenburgh H, et al. Muscle Nerve. 2008;37:438–47.

Gene Therapy in Cartilage Repair and Regeneration

5

Sheri Wang and Bing Wang

5.1 Introduction

The regeneration of diseased or damaged cartilage tissues remains a significant clinical challenge. Characterized by progressive inflammation and the degeneration of articular cartilage tissue, cartilage diseases and injuries eventually compromise a patient's mobility, resulting in joint damage, arthritis, and disability. Current medical treatments targeting cartilage are effective at reducing pain but ineffective at reversing the course of the disease, which is rooted in low cellularity, limited vascularity, and chronic inflammation. Traditional pharmacotherapy delivers desired proteins directly to the site of injury through injections; however, pharmacotherapy has a limited half-life and is quickly degraded and cleared by the body. As a result, therapeutic concentrations of drugs are difficult to maintain at the site of injury. Therefore, the use of growth factors and anti-inflammatory agents can only temporarily arrest the course of disease and injury, often requiring long-term treatment with systemic immunosuppressive therapy, such as those used in the treatment of arthritis.

As opposed to using protein agents, gene therapy delivers therapeutic genes using nonpathogenic viral vectors through an in vivo or ex vivo gene transfer approach, providing novel treatments for cartilage degenerative diseases. As depicted in Fig. 5.1, current recombinant adeno-associated viral (AAV) vector-based gene therapy strategies for the regeneration of articular cartilage tissue include (1) in vivo therapy (direct rAAV injection into the site of injury), (2) ex vivo therapy (indirect, in vitro infection of cells which will later be introduced to the site of injury), and (3) biomaterial-mediated gene transfer (scaffolding biomaterial is

S. Wang
University of Pittsburgh, Pittsburgh, PA, USA

B. Wang (✉)
Department of Orthopaedic Surgery, University of Pittsburgh, Pittsburgh, PA, USA
e-mail: bingwang@pitt.edu

G. Giatsidis (ed.), *Gene Therapy in Reconstructive and Regenerative Surgery*,
https://doi.org/10.1007/978-3-319-78957-6_5

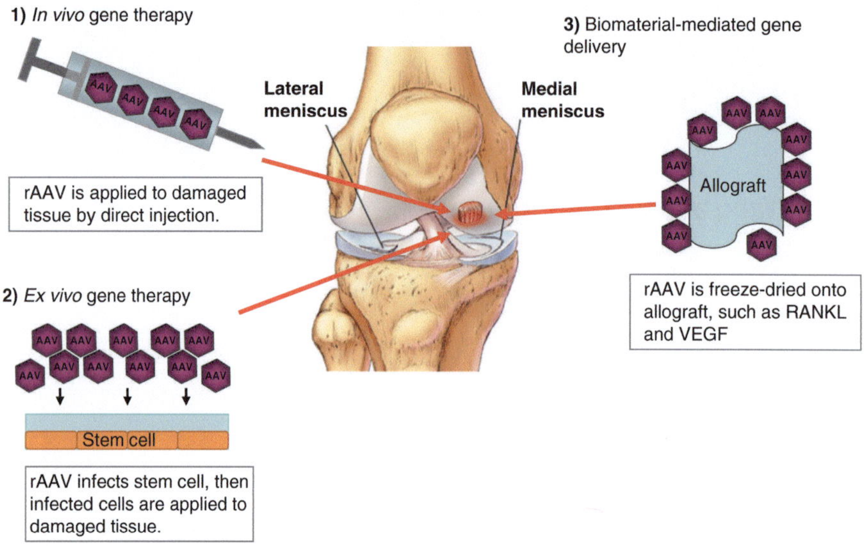

Fig. 5.1 AAV-based gene therapy strategies for the reconstruction of damaged articular cartilage tissue

impregnated with rAAV vector). By using the above strategies, AAV is growing as the preferred choice for many human applications [1], including for articular cartilage disease and injury [2, 3]. Unlike conventional therapies such as steroids, gene therapy, theoretically, could ultimately result in a cure for articular cartilage disease and injury. Gene therapy is a promising therapeutic option for articular cartilage tissue regeneration.

In this chapter, we will discuss AAV-based in vivo and ex vivo gene therapy approaches that are currently being evaluated. A special focus will be given to AAV delivery strategies and clinical applications, such as arthritis. The goal of this chapter is to provide the readers with a broad background of gene therapy for the cartilage and an introduction to relevant topics in AAV mediated cartilage repair, including gene expression of growth factors for cartilage regeneration and repair, as well as expressing inhibitors for reducing inflammation.

5.2 Development of AAV Vectors

The AAV vector is derived from nonpathogenic and non-replicative parvoviruses [4] with a small (20 nm) regular icosahedral viral particle [5]. The wild-type AAV viral genome consists of either positive- or negative-sensed, single-stranded DNA around 4.7 kilo-base pairs (kb) long with inverted terminal repeats (ITRs) at both ends and two open reading frames (ORFs): Rep and Cap. Recombinant AAV vectors have all viral genes removed and only retain the 145 base pair (bp) ITR flanking the foreign gene(s). The maximal capacity for the foreign gene expression cassette in a single-stranded AAV vector, including the sizes of promoter, gene, and polyadenylation (polyA) tail, should be <4.7 kb. AAV vectors have the distinct, advantageous biological property of being capable of infecting a wide range of host cell types, including dividing and nondividing cells. This quality makes AAV an ideal vector for treating injured tissues, such as articular cartilage, which do not undergo rapid repair after injury. Although wild-type AAV viruses can perform a site-specific, Rep 78-mediated integration at AAVS1 on chromosome 19 [6], the genome of an AAV vector often remains in episomal form [7] in the nucleus of host cells due to the elimination of all the viral genes (Rep and Cap). Hence, the risk of low efficiency, immune response, or potential insertional mutagenesis caused by AAV vector is very low compared to the other viral vectors, such as retro- or lentiviral vectors [8, 9]. Thus far, AAV-based gene therapy represents one of the most promising approaches to aid in the repair and regeneration of articular cartilage tissue.

AAV-mediated gene therapy strategies have been developed for numerous musculoskeletal diseases and injuries, including arthritis, several of which are currently in or close to the clinical phase. The benefits of utilizing AAV include superior long-term gene transfer efficiency for the treatment of cartilage disease and injury, minimal immune response to the virus, and low toxicity, which is a problem often associated with other viral vectors such as retrovirus and adenovirus. AAV-mediated local gene delivery also has the advantage of being able to overcome systemic side effects while providing long-term therapeutic outcomes following a single vector injection. Moreover, the long-term pharmacologically regulated delivery of a desired therapeutic agent following AAV-mediated gene transfer [10, 11] provides for optimal safety and efficacy that is generally required for enhancing tissue repair in regenerative medicine. As it progresses into clinical trials for the treatment of cartilage degeneration-related arthritis, AAV-mediated gene therapy demonstrates promise as an efficacious, safe, and convenient treatment.

A major challenge in the field is to ensure the efficiency of gene transfer in the target tissue in large animal models and human clinical trials, especially following systemic AAV vector delivery. We have been able to enhance gene delivery used for in vivo directly to body or ex vivo engineered stem cell through modifications of viral particles and genomes. Further advancements can be made by studying intracellular trafficking of AAV vectors. Since the birth of AAV-based gene therapy, research has been focused on the development of novel AAV vector systems and the optimization of existing ones. Based on our biological knowledge of AAV, two strategies have been used to develop highly efficient and rapidly expressed AAV vectors through (1) the modification of particles to increase the affinity of AAV to host cells at the beginning of the intracellular trafficking process and (2) the modification of the viral genome to bypass the rate-limiting conversion step of ssDNA to dsDNA.

Twelve distinct serotypes (AAV1–AAV12) have been identified so far, each containing a different capsid amino acid sequence, with over 100 variants in humans and nonhuman primates [12, 13]. The 12th AAV serotype was discovered just recently, in 2008. Current research indicates that the considerable diversity in different cell types and target tissues of wild-type AAV virus transduction can be attributed to the unique tropism of different AAV serotypes [14]. Therefore, highly efficient delivery of transgene(s) in the musculoskeletal system has been achieved by choosing the appropriate serotype and a correct delivery approach, giving AAV enormous potential as a vector for human gene therapy. As a result, the mutant AAV particles are characterized by enhanced receptor binding and virus uptake by either altering capsid plasticity [15] or by the inclusion of a tissue-targeting peptide. Well-designed AAV capsid mutants pave the way to achieving efficient transduction in musculoskeletal tissues.

Moreover, the low transduction of natural AAV serotypes confines their action to genetically modify stem cells [16]. More recently, "designer" AAV vector mutants have been created from several diverse AAV capsid libraries including randomly mutagenized [17], DNA shuffled [18], and random peptide insertion variants [19] of a selected peptide sequence on the surface of the threefold spike within the heparin-binding site [20]. These "designer" vectors have been used for efficient gene delivery and gene targeting within pluripotent stem cells isolated from rat, mouse, and human. Such an approach is promising not only for basic stem cell biology but also for clinical applications.

In summary, the standard rule of designing ideal AAV vectors depends on the length of the transferred gene's cDNA and the goal of targeting specific tissues. More specifically, (1) the self-complementary AAV vectors (scAAV) with a typically faster and higher gene expression is suitable for packaging cDNA <1.2 kb, which includes most growth factors; (2) a regulated AAV delivery system is likely best used for the healing of cartilage defects, as well as for reducing inflammation; (3) the appropriate serotype and tissue-specific promoter can enhance gene expression efficiency and reduce the risk of side effects; and (4) for the genetic modification of stem cells, enhanced gene delivery can be achieved by the development of AAV variants with the specific insertion of the viral genome into the host chromosome, as well as the combination of AAV variants with biomaterials.

5.3 AAV-Based In Vivo Gene Therapy

Over the past decade, gene transfer vehicles based on engineered AAV have consistently emerged as the therapeutic vector of choice for a broad range of genetic and chronic diseases in musculoskeletal system, including arthritis and articular cartilage repair [21–27]. We know chondrocytes in damaged articular cartilage play a key role during healing, but the most appropriate serotype of AAV for gene delivery in human chondrocytes has yet to be determined. As shown in Fig. 5.2, the green fluorescent protein (GFP) was used to examine the efficiency of gene transfer, and we found that AAV6 showed high level of GFP expression in primary chondrocytes 24 h post-infection.

In addition to the use of the correct serotype of AAV for in vivo gene delivery, the choice of a route of administration can achieve highly efficient expression of transgenes in diseased or damaged tissues. The local treatment is preferred over treating cartilage tissue and inflammatory joint disorders through intraarticular injection with gene vectors, in order to enhance cartilage tissue healing and reduce chronic inflammation in damaged joints. The advantages of local gene delivery include (1)

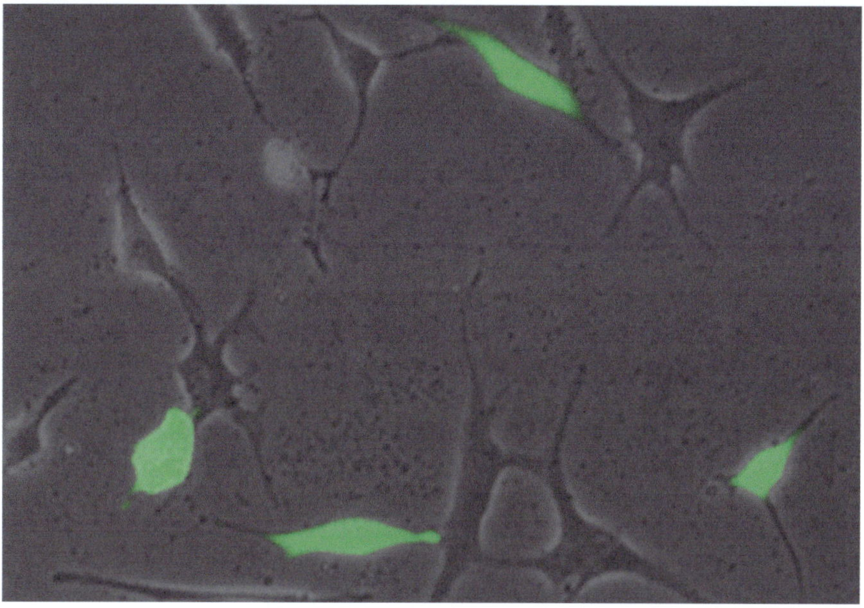

Fig. 5.2 The highly efficient GFP gene expression mediated by AAV6 was observed in human primary chondrocytes. Magnification 40×

long-term therapeutic effects following a single treatment without considering systemic side effects and (2) regulated expression of a gene through an inducible AAV system. By using the local gene delivery strategy, AAV vectors have been applied to arthritis and articular cartilage repair via the expression of anti-inflammatory growth factors and cytokines.

Intraarticular administration of AAV vectors to block chronic inflammation and reverse the course of cartilage destruction in affected joints has been developed as a novel treatment for chronic inflammatory conditions such as rheumatoid arthritis (RA), providing an alternative method to traditional drug delivery. Gene therapy for RA targets compounds affecting inflammation or articular destruction. Inhibition of pro-inflammatory cytokines was first developed as a novel strategy for gene therapy for chronic arthritis, including TNF-α or IL-1 blocking agents [22, 28, 29] (such as anti-TNF-α monoclonal antibodies, soluble TNF-α receptor, type II soluble receptor of IL-1, IL-1 receptor antagonist) and anti-inflammatory interleukins [30–33] (such as IL-4, IL-10). Moreover, the classical nuclear factor κB (NF-κB) signaling pathway is known to be pivotal for regulating synthesis and activity of inflammatory cytokines (TNF-α, IL-1β) and also for several other mediators involved in RA and OA, including cyclooxygenase-2 (COX-2), inducible nitric oxide synthase (iNOS), and MMP-1 [34, 35]. Many therapeutic strategies have focused on inhibitor of NF-κB kinase β (IKKβ), which plays the most prominent role in the activation of NF-κB by pro-inflammatory cytokines in the pathological progression of RA [36]. In the rat model, a dominant-negative IKKβ K → M (IKKβ-dn) mutant carried by an AAV vector through a local intra-articular injection significantly ameliorated the severity of adjuvant arthritis according to clinical parameters (paw thickness) and inflammation parameters (TNF, IL-1, IL-10, and MMP-1) [23]. AAV-based, locally sustained suppression of inflammation did not lead to observable systemic side effects.

In addition to the blockage of inflammation, additional therapeutic strategies for chronic arthritis include reversing the course of cartilage destruction by promoting anabolic activities of articular chondrocytes and healing of articular cartilage in the damaged joint. AAV-mediated delivery of growth factors is a viable therapeutic strategy for chronic arthritis [37, 38]. In this case, AAV would be used to reverse the course of cartilage destruction by promoting the anabolic activities of articular chondrocytes and healing of articular cartilage in damaged joints. AAV-mediated fibroblast growth factor (FGF-2) was not only sufficient to stimulate chondrocyte proliferation but also significantly improved the overall repair, filling, architecture, and cell morphology of osteochondral defects in rabbit knee joints [24]. Furthermore, Hiraide et al. investigated the intra-articular administration of AAV vector carrying basic fibroblast growth factor (bFGF) and found it could enhance regeneration of cartilage and result in a significant overall repair of osteochondral defects and articular cartilage in a rabbit model [25]. Another growth factor, transforming growth factor (TGFβ1), has also been proven to stimulate cartilage anabolism in normal and osteoarthritic human chondrocytes by AAV vector delivery [26]. More recently, Izal et al. reported that AAV-mediated insulin-like growth factor-1 (IGF-1) gene therapy was able to protect cartilage in a rat model of joint damage but not sufficient for the repair of seriously damaged cartilage [27].

To ensure safety in clinical application, AAV-mediated expression of immunosuppressive cytokines or growth factors for the treatment of chronic inflammatory disorders in musculoskeletal system must not maintain long-term transgene expression. Otherwise, permanently expressing a transgene in locally damaged tissues could lead to severe side effects. Therefore, therapeutic gene expression should be maximal during the flare-up of the disease and switched off during remission of the disease. In this case, AAV-mediated regulated expression is particularly important because it is capable of turning off transgene expression following the healing of the disease, avoiding potential toxicity. Based on a tetracycline-inducible transactivator, currently designed inducible AAV vector systems allow the pharmacological regulation of transgene expression for cartilage regeneration, such as luciferase reporter in rat knees [39] and interleukin-10 in the arthritic CIA mouse [33]. In addition, the regulated AAV gene delivery systems using on-and-off regulation of an inducible promoter are more suitable for cartilage gene therapy because they offer transient expression of a desired protein for cartilage tissue repair and regeneration, which avoids the potential toxicity caused by long-term transgene overexpression.

5.4 AAV-Based Ex Vivo Gene Therapy

Adult stem cells isolated from different sources including the bone marrow, adipose tissue, skeletal muscle, or tendon possess the capacity for self-renewal and multilineage differentiation. These cells can be used to regenerate tissues of the musculoskeletal system in the context of genetic disorders, chronic inflammation-related degenerative diseases, as well as tissue injuries [40–44]. Unlike embryonic stem (ES) cells, adult stem cells have the advantages of low risk of tumorigenesis, immune rejection, and few ethical concerns [45]. The US Food and Drug Administration (FDA) approved stem cells for the treatment of fatal diseases in clinical trials. Adult stem cell-based therapy has potentially opened the door for various new treatments [46].

However, there are challenges that come with using adult stem cells: most adult stem cells are pre-specialized, with a low capacity for proliferation and differentiation to certain tissues such as articular cartilage. Furthermore, gene correction or targeted integration to modify adult stem cells is necessary for the treatment of genetic diseases through autologous cell therapy and cell-based therapies that enhance tissue regeneration through the overexpression of growth factors. During the last decade, therapeutic benefits were limited to retro- and lentiviral vectors. However, due to the uncontrolled potential for insertional mutagenesis in vivo, safer methods for delivering therapeutic genes are being sought. AAV vector-based cell therapy, called ex vivo gene therapy, involves AAV transduction of an appropriate factor in cultured cells followed by the administration of such genetic-engineered cells into the target tissue. As a result, the desired protein, such as a growth factor, will be expressed and secreted from the genetically modified cells in high concentrations at the site of injection to improve cartilage repair.

So far, it remains a challenge to regulate the culture environment of stem cell proliferation and chondrogenic differentiation in naïve cell-based regeneration of damaged or diseased articular cartilage. Gene transfer combined with stem cell therapy is believed to be the future of cartilage regeneration. The transduction of stem cells with potent therapeutic genes, such as growth factors, can further enhance their therapeutic capability prior to their application for regenerative medicine. In fact, stem cells from different sources would be excellent targets for gene transfer, as, in principle, a single corrected or modified cell could provide lifelong repopulation of the host with modified cells and could benefit multiple lineages. The combination of gene and cell therapies using genetically modified cells offers several unique advantages over direct gene transfer into the body and cell therapy alone, which involves administration of cells that have not been genetically modified.

By using ex vivo gene transfer, the desired protein or growth factor will be expressed and secreted from the genetically modified cells at a high level at the site of injection, improving tissue repair. Using this approach, Yokoo et al. isolated primary rabbit chondrocytes followed by transduction of AAV-lacZ or AAV-bFGF [47]. For example, AAV2-based bFGF gene-transduced rabbit chondrocytes were embedded in a collagen gel and transplanted into a full-thickness defect in the articular cartilage of the patellar groove of a rabbit, resulting in repair of the articular cartilage defect. The genetically modified cells could be found at the transplant sites for 4 weeks and improved the healing of cartilage defects [24]. Another ex vivo therapy using AAV-TGF-β1 demonstrates that genetic engineering of HuMSCs improves cartilage repair in osteochondral defects of athymic rats [48]. In the latter, ultraviolet (UV) light-activated gene transduction (LAGT) combined with AAV is also a promising way to facilitate TGF-β1 gene induction in HuMSCs for cartilage and bone repair by enhancing the efficiency of the gene transduction. More recently, Frisch et al. demonstrated that the AAV vector is also a safe and efficient delivery system for the genetic modification of human bone marrow-derived mesenchymal stem cells via ex vivo IGF-1gene transfer [49]. The implanted mesenchymal stem cells were able to release therapeutic levels of IGF-1 to sustained mesenchymal stem cell chondrogenesis for cartilage regeneration, indicating that a novel treatment for articular cartilage degenerative diseases is in the near future [49, 50]. Moreover, the ability to regulate expression of target genes in progenitor cells is an important tool for their study and application toward gene and cell therapy [51]. The AAV inducible system provides pharmacological controlled expression of chondrogenic growth and differentiation factors in progenitor cells, improving efficacy and safety during tissue healing.

5.5 AAV-Based Clinical Trials and Challenges

Consequently, the manufacture of clinical-grade AAV vectors is very important for the delivery of therapeutic genes utilized in these trials. Therefore, the production of AAV vectors must be guided by Good Manufacturing Practices (GMP) conditions so that the product is of a known, consistent quality, quantity, and potency. However, the manufacture of clinical-grade AAV vectors remains significantly challenging, with little consensus as to "best practices" for the manufacturing process or clearly defined biochemical assays for ensuring final product potency. Currently, a vector expressing a codon-optimized therapeutic gene driven by a tissue's specific promoter has been manufactured on a large-scale GMP production and gained regulatory approval in different countries such as the USA, UK, and China [52]. Accordingly, the efficiency, safety, and convenience associated with AAV-mediated gene therapy have led to the initiation of clinical trials for the treatment of RA, by reducing inflammation and enhancing cartilage regeneration, respectively.

Gene therapy approaches for treating autoimmune disorders such as RA have shown significant efficacy in numerous animal models. Using intra-articular injection of an AAV vector expressing a fusion protein (TNFR:Fc) combining the extracellular domain of human TNF-α receptor type II and the Fc domain of Ig-G1 was impressive at reducing inflammation and inhibiting bone destruction in a rat RA model. Based on this strategy, the first AAV-based human gene therapy clinical trial for the treatment of RA was initiated in 2005 by Targeted Genetics Inc. in Seattle Washington. These Phase I/II trials are utilizing an AAV2 vector and intra-articular delivery of the TNF-α antagonist, etanercept, to inhibit TNF-α, reducing inflammation [53]. The Phase I trial was accomplished without any serious side effects [54]. In the Phase II trial (OBA protocol: 0504-705), 127 patients in 20 research centers throughout the USA were enrolled to receive the first injection, of which 74 received the second injection of AAV2-CMV-TNFR:Fc. Unfortunately, a 36-year-old patient with RA died 22 days after receiving a second dose of AAV administration [55, 56]. The patient's death was due to disseminated histoplasmosis that was complicated by retroperitoneal bleeding after long-term treatment with systemic immunosuppressive therapy for RA. This death was widely publicized in the press, and the trial was put on hold while the authorities investigated the matter [57]. After a more than 2-year investigation, the evidence demonstrated that the AAV vector was inculpable. The FDA allowed the trials to continue with a modification to the protocol, and the trial was completed without further incident. Although this incident was not directly associated with AAV treatment, the selection and monitoring of patients before and after AAV injection are critical. Whether there will be further clinical trials of AAV2-TNFR:Fc, or indeed of any other gene therapy, for RA in the near future is unknown [58]. AAV-based in vivo treatment for arthritis gene therapy is a relatively new approach in medicine, and its techniques still require refinement.

Overall the utilization of AAV vectors in preclinical and clinical gene transfer applications is rising. In the future, precautionary monitoring plans should be taken for the patients in the trials, such as (1) checking antibody of capsid, (2) developing assay to quantify transgene product, and (3) screening for signs of active infection.

Conclusion

During the past decade, advantages such as high efficiency and minimal side effects have demonstrated that AAV vectors are particularly useful in initiating and accelerating the repair of cartilage, as well as the treatment of chronic inflammation [59]. The use of the appropriate AAV serotype allows for the transferring therapeutic genes to targeted tissues through local administration. Cellular and gene therapies are a highly complementary pairing. The effective approaches include direct, in vivo gene transfer of a therapeutic gene and the ex vivo method involving genetically modified stem cells, as well as combination with biomaterials. To further improve the gene delivery properties of the current AAV2-based vector for human clinical trials, scientists are applying for a directed evolutionary approach to create AAV variants for enhancing gene delivery efficiency that can reduce immune response or cytotoxic T-lymphocyte reaction, as well as enhance gene delivery in human stem cells. Finally, a well-designed monitoring plan to prevent infectious complications is very important for the clinical trials studying chronic inflammatory disorders.

Acknowledgments We gratefully acknowledge Ms. Anna T. Dzuricky, University of Pittsburgh Swanson School of Engineering, for reviewing and editing this manuscript.

References

1. Warrington KH Jr, Herzog RW. Treatment of human disease by adeno-associated viral gene transfer. Hum Genet. 2006;119(6):571–603.
2. Zhang X, Godbey WT. Viral vectors for gene delivery in tissue engineering. Adv Drug Deliv Rev. 2006;58(4):515–34.
3. Dai J, Rabie AB. The use of recombinant adeno-associated virus for skeletal gene therapy. Orthod Craniofac Res. 2007;10(1):1–14.

4. Atchison RW, Casto BC, Hammon WM. Adenovirus-associated defective virus particles. Science. 1965;149:754–6.
5. Xie Q, Bu W, Bhatia S, Hare J, Somasundaram T, Azzi A, et al. The atomic structure of adeno-associated virus (AAV-2), a vector for human gene therapy. Proc Natl Acad Sci U S A. 2002;99(16):10405–10.
6. Weitzman MD, Kyostio SR, Kotin RM, Owens RA. Adeno-associated virus (AAV) rep proteins mediate complex formation between AAV DNA and its integration site in human DNA. Proc Natl Acad Sci U S A. 1994;91(13):5808–12.
7. Flotte TR, Afione SA, Zeitlin PL. Adeno-associated virus vector gene expression occurs in nondividing cells in the absence of vector DNA integration. Am J Respir Cell Mol Biol. 1994;11(5):517–21.
8. Cone RD, Mulligan RC. High-efficiency gene transfer into mammalian cells: generation of helper-free recombinant retrovirus with broad mammalian host range. Proc Natl Acad Sci U S A. 1984;81(20):6349–53.
9. Luk KD, Chen Y, Cheung KM, Kung HF, Lu WW, Leong JC. Adeno-associated virus-mediated bone morphogenetic protein-4 gene therapy for in vivo bone formation. Biochem Biophys Res Commun. 2003;308(3):636–45.
10. Ye X, Rivera VM, Zoltick P, Cerasoli F Jr, Schnell MA, Gao G, et al. Regulated delivery of therapeutic proteins after in vivo somatic cell gene transfer. Science. 1999;283(5398):88–91.
11. Rivera VM, Gao GP, Grant RL, Schnell MA, Zoltick PW, Rozamus LW, et al. Long-term pharmacologically regulated expression of erythropoietin in primates following AAV-mediated gene transfer. Blood. 2005;105(4):1424–30. https://doi.org/10.1182/blood-2004-06-2501.
12. Grimm D, Kay MA. From virus evolution to vector revolution: use of naturally occurring serotypes of adeno-associated virus (AAV) as novel vectors for human gene therapy. Curr Gene Ther. 2003;3(4):281–304.
13. Mori S, Wang L, Takeuchi T, Kanda T. Two novel adeno-associated viruses from cynomolgus monkey: pseudotyping characterization of capsid protein. Virology. 2004;330(2):375–83.
14. Ding W, Zhang L, Yan Z, Engelhardt JF. Intracellular trafficking of adeno-associated viral vectors. Gene Ther. 2005;12(11):873–80.
15. Schaffer DV, Maheshri N. Directed evolution of AAV mutants for enhanced gene delivery. Conf Proc IEEE Eng Med Biol Soc. 2004;5:3520–3.
16. Asuri P, Bartel MA, Vazin T, Jang JH, Wong TB, Schaffer DV. Directed evolution of adeno-associated virus for enhanced gene delivery and gene targeting in human pluripotent stem cells. Mol Ther. 2012;20(2):329–38. https://doi.org/10.1038/mt.2011.255.
17. Grimm D, Lee JS, Wang L, Desai T, Akache B, Storm TA, et al. In vitro and in vivo gene therapy vector evolution via multispecies interbreeding and retargeting of adeno-associated viruses. J Virol. 2008;82(12):5887–911. https://doi.org/10.1128/JVI.00254-08.
18. Koerber JT, Jang JH, Schaffer DV. DNA shuffling of adeno-associated virus yields functionally diverse viral progeny. Mol Ther. 2008;16(10):1703–9. https://doi.org/10.1038/mt.2008.167.
19. Adachi K, Nakai H. A new recombinant adeno-associated virus (Aav)-based random peptide display library system: infection-defective Aav1.9-3 as a novel detargeted platform for vector evolution. Gene Ther Regul. 2010;5(1):31–55. https://doi.org/10.1142/S1568558610000197.
20. Jang JH, Koerber JT, Kim JS, Asuri P, Vazin T, Bartel M, et al. An evolved adeno-associated viral variant enhances gene delivery and gene targeting in neural stem cells. Mol Ther. 2011;19(4):667–75. https://doi.org/10.1038/mt.2010.287.
21. Zhang HG, Xie J, Yang P, Wang Y, Xu L, Liu D, et al. Adeno-associated virus production of soluble tumor necrosis factor receptor neutralizes tumor necrosis factor alpha and reduces arthritis. Hum Gene Ther. 2000;11(17):2431–42.
22. Pan RY, Chen SL, Xiao X, Liu DW, Peng HJ, Tsao YP. Therapy and prevention of arthritis by recombinant adeno-associated virus vector with delivery of interleukin-1 receptor antagonist. Arthritis Rheum. 2000;43(2):289–97.
23. Tak PP, Gerlag DM, Aupperle KR, van de Geest DA, Overbeek M, Bennett BL, et al. Inhibitor of nuclear factor kappaB kinase beta is a key regulator of synovial inflammation. Arthritis Rheum. 2001;44(8):1897–907.

24. Cucchiarini M, Madry H, Ma C, Thurn T, Zurakowski D, Menger MD, et al. Improved tissue repair in articular cartilage defects in vivo by rAAV-mediated overexpression of human fibroblast growth factor 2. Mol Ther. 2005;12(2):229–38.
25. Hiraide A, Yokoo N, Xin KQ, Okuda K, Mizukami H, Ozawa K, et al. Repair of articular cartilage defect by intraarticular administration of basic fibroblast growth factor gene, using adeno-associated virus vector. Hum Gene Ther. 2005;16(12):1413–21.
26. Ulrich-Vinther M, Stengaard C, Schwarz EM, Goldring MB, Soballe K. Adeno-associated vector mediated gene transfer of transforming growth factor-beta1 to normal and osteoarthritic human chondrocytes stimulates cartilage anabolism. Eur Cell Mater. 2005;10:40–50.
27. Izal I, Acosta CA, Ripalda P, Zaratiegui M, Ruiz J, Forriol F. IGF-1 gene therapy to protect articular cartilage in a rat model of joint damage. Arch Orthop Trauma Surg. 2008;128(2):239–47.
28. Bakker AC, Joosten LA, Arntz OJ, Helsen MM, Bendele AM, van de Loo FA, et al. Prevention of murine collagen-induced arthritis in the knee and ipsilateral paw by local expression of human interleukin-1 receptor antagonist protein in the knee. Arthritis Rheum. 1997;40(5):893–900.
29. Kim JM, Jeong JG, Ho SH, Hahn W, Park EJ, Kim S, et al. Protection against collagen-induced arthritis by intramuscular gene therapy with an expression plasmid for the interleukin-1 receptor antagonist. Gene Ther. 2003;10(18):1543–50.
30. Cottard V, Mulleman D, Bouille P, Mezzina M, Boissier MC, Bessis N. Adeno-associated virus-mediated delivery of IL-4 prevents collagen-induced arthritis. Gene Ther. 2000; 7(22):1930–9.
31. Goater J, Muller R, Kollias G, Firestein GS, Sanz I, O'Keefe RJ, et al. Empirical advantages of adeno associated viral vectors in vivo gene therapy for arthritis. J Rheumatol. 2000;27(4):983–9.
32. Saidenberg-Kermanac'h N, Bessis N, Deleuze V, Bloquel C, Bureau M, Scherman D, et al. Efficacy of interleukin-10 gene electrotransfer into skeletal muscle in mice with collagen-induced arthritis. J Gene Med. 2003;5(2):164–71.
33. Apparailly F, Millet V, Noel D, Jacquet C, Sany J, Jorgensen C. Tetracycline-inducible interleukin-10 gene transfer mediated by an adeno-associated virus: application to experimental arthritis. Hum Gene Ther. 2002;13(10):1179–88.
34. Handel ML, Girgis L. Transcription factors. Best Pract Res Clin Gastroenterol. 2001;15(5):657–75.
35. Roshak AK, Callahan JF, Blake SM. Small-molecule inhibitors of NF-kappaB for the treatment of inflammatory joint disease. Curr Opin Pharmacol. 2002;2(3):316–21.
36. Tas SW, Adriaansen J, Hajji N, Bakker AC, Firestein GS, Vervoordeldonk MJ, et al. Amelioration of arthritis by intraarticular dominant negative Ikk beta gene therapy using adeno-associated virus type 5. Hum Gene Ther. 2006;17(8):821–32.
37. Firestein GS, Zvaifler NJ. Anticytokine therapy in rheumatoid arthritis. N Engl J Med. 1997;337(3):195–7.
38. Bessis N, Doucet C, Cottard V, Douar AM, Firat H, Jorgensen C, et al. Gene therapy for rheumatoid arthritis. J Gene Med. 2002;4(6):581–91.
39. Payne KA, Lee HH, Haleem AM, Martins C, Yuan Z, Qiao C, et al. Single intra-articular injection of adeno-associated virus results in stable and controllable in vivo transgene expression in normal rat knees. Osteoarthr Cartil. 2011;19(8):1058–65. https://doi.org/10.1016/j.joca.2011.04.009.
40. Peng H, Huard J. Muscle-derived stem cells for musculoskeletal tissue regeneration and repair. Transpl Immunol. 2004;12(3–4):311–9. https://doi.org/10.1016/j.trim.2003.12.009.
41. Zheng B, Cao B, Crisan M, Sun B, Li G, Logar A, et al. Prospective identification of myogenic endothelial cells in human skeletal muscle. Nat Biotechnol. 2007;25(9):1025–34.
42. Salgado AJ, Oliveira JT, Pedro AJ, Reis RL. Adult stem cells in bone and cartilage tissue engineering. Curr Stem Cell Res Ther. 2006;1(3):345–64.
43. Bi Y, Ehirchiou D, Kilts TM, Inkson CA, Embree MC, Sonoyama W, et al. Identification of tendon stem/progenitor cells and the role of the extracellular matrix in their niche. Nat Med. 2007;13(10):1219–27. https://doi.org/10.1038/nm1630.

44. Korbling M, Estrov Z. Adult stem cells for tissue repair - a new therapeutic concept? N Engl J Med. 2003;349(6):570–82. https://doi.org/10.1056/NEJMra022361.
45. Hall VJ, Stojkovic P, Stojkovic M. Using therapeutic cloning to fight human disease: a conundrum or reality? Stem Cells. 2006;24(7):1628–37. https://doi.org/10.1634/stemcells.2005-0592.
46. Kimelman N, Pelled G, Helm GA, Huard J, Schwarz EM, Gazit D. Review: gene- and stem cell-based therapeutics for bone regeneration and repair. Tissue Eng. 2007;13(6):1135–50.
47. Yokoo N, Saito T, Uesugi M, Kobayashi N, Xin KQ, Okuda K, et al. Repair of articular cartilage defect by autologous transplantation of basic fibroblast growth factor gene-transduced chondrocytes with adeno-associated virus vector. Arthritis Rheum. 2005;52(1):164–70. https://doi.org/10.1002/art.20739.
48. Pagnotto MR, Wang Z, Karpie JC, Ferretti M, Xiao X, Chu CR. Adeno-associated viral gene transfer of transforming growth factor-beta1 to human mesenchymal stem cells improves cartilage repair. Gene Ther. 2007;14(10):804–13.
49. Frisch J, Venkatesan JK, Rey-Rico A, Schmitt G, Madry H, Cucchiarini M. Influence of insulin-like growth factor I overexpression via recombinant adeno-associated vector gene transfer upon the biological activities and differentiation potential of human bone marrow-derived mesenchymal stem cells. Stem Cell Res Ther. 2014;5(4):103. https://doi.org/10.1186/scrt491.
50. Tang Y, Wang B. Gene- and stem cell-based therapeutics for cartilage regeneration and repair. Stem Cell Res Ther. 2015;6:78. https://doi.org/10.1186/s13287-015-0058-5.
51. Moutsatsos IK, Turgeman G, Zhou S, Kurkalli BG, Pelled G, Tzur L, et al. Exogenously regulated stem cell-mediated gene therapy for bone regeneration. Mol Ther. 2001;3(4):449–61. https://doi.org/10.1006/mthe.2001.0291.
52. Allay JA, Sleep S, Long S, Tillman DM, Clark R, Carney G, et al. Good manufacturing practice production of self-complementary serotype 8 adeno-associated viral vector for a hemophilia B clinical trial. Hum Gene Ther. 2011;22(5):595–604. https://doi.org/10.1089/hum.2010.202.
53. Anklesaria P, Heald EA, Mease JP. Intra-articular administration of a recombinant adeno-associated vector containing a TNF antagonist gene was safe, well tolerated and demonstrated trend in clinical response in subjects with inflammatory arthritis. In: 11th American Society of Gene Therapy annual meeting; May 28–June 1; Boston, MA, US; 2008.
54. Mease PJ, Hobbs K, Chalmers A, El-Gabalawy H, Bookman A, Keystone E, et al. Local delivery of a recombinant adenoassociated vector containing a tumour necrosis factor alpha antagonist gene in inflammatory arthritis: a phase 1 dose-escalation safety and tolerability study. Ann Rheum Dis. 2009;68(8):1247–54. https://doi.org/10.1136/ard.2008.089375.
55. Kaiser J. Clinical research. Death prompts a review of gene therapy vector. Science. 2007;317(5838):580. https://doi.org/10.1126/science.317.5838.580.
56. Evans CH, Ghivizzani SC, Robbins PD. Arthritis gene therapy's first death. Arthritis Res Ther. 2008;10(3):110. https://doi.org/10.1186/ar2411.
57. Frank KM, Hogarth DK, Miller JL, Mandal S, Mease PJ, Samulski RJ, et al. Investigation of the cause of death in a gene-therapy trial. N Engl J Med. 2009;361(2):161–9.
58. Evans CH, Ghivizzani SC, Robbins PD. Getting arthritis gene therapy into the clinic. Nat Rev Rheumatol. 2011;7(4):244–9. https://doi.org/10.1038/nrrheum.2010.193.
59. Evans CH, Ghivizzani SC, Gouze E, Rediske JJ, Schwarz EM, Robbins PD. The 3rd International Meeting on Gene Therapy in Rheumatology and Orthopaedics. Arthritis Res Ther. 2005;7(6):273–8.

6 Gene Therapy in Tendon Repair and Regeneration

Raman Mehrzad, Xiao Tian Wang, You Lang Zhou,
Ya Fang Wu, and Paul Y. Liu

Abbreviations

AAV	Adeno-associated virus
bFGF	Basic fibroblast growth factor
PDGF-B	Platelet-derived growth factor-B
VEGF	Vascular endothelial growth factor

6.1 Tendon Injuries

Tendon and ligament injuries constitute one of the most common traumas to the human body, with tendon injuries to the hand and wrist occurring in hundreds of thousands of people annually in this country alone [1, 2]. Serious tendon lacerations result in millions of lost days from work each year. With >100,000 injuries/year, at least 3 months out of work/patient, and a re-rupture rate after surgical repairing (with subsequent second operation) around 10–20%, the estimated cost of tendon injuries of the hand in the USA is >$1.2 billion annually. In fact, injuries in tendons are ranked first in the order of most expensive injury types [3], and significant permanent disability from incomplete rehabilitation is all too often the final result. Tendon injuries can result from trauma, overuse, or age-related degeneration from

R. Mehrzad · X. T. Wang · P. Y. Liu (✉)
Department of Plastic Surgery, Rhode Island Hospital, The Warren Alpert Medical School of Brown University, Providence, RI, USA
e-mail: xwang3@Lifespan.org; pliu@lifespan.org

Y. L. Zhou · Y. F. Wu
Department of Hand Surgery, The Hand Surgery Research Center, Affiliated Hospital of Nantong University, Nantong, Jiangsu, China
e-mail: yafang.wu@ntu.edu.cn

G. Giatsidis (ed.), *Gene Therapy in Reconstructive and Regenerative Surgery*,
https://doi.org/10.1007/978-3-319-78957-6_6

work, daily life, and sports activities. Since physical exercise is frequently a major part of many professions' daily schedule such as military personnel, professional athletes, etc., they tend to suffer a higher incidence of tendon injuries than most others and are in high demand of proper healing [4–7].

Tendons are tissues that connect muscles to the bone. A body part, such as a finger, is able to move due to the contraction of a muscle, in which the tendon pulls the bone. Due to the tendon's close proximity to the surface of the skin, a deep cut is likely to hit a tendon and cause a partial or complete division. A completely torn tendon will make it impossible to bend the joint(s). Because of the large amount of resting tension presented to the tendon, the tendon ends will be pulled apart; thus, the majority of tendon injuries require surgical treatment [8].

6.2 Phases of Tendon Healing

In general, there are three major phases to tendon healing [9, 10]:

1. Inflammatory stage—This stage encompasses the first few days after injury, with formation of a fibrin clot, which is then invaded by polymorphonuclear leukocytes and other inflammatory cells. The function of these cells is to remove debris and damage and to promote angiogenesis and proliferation of fibroblasts, by the release of growth factors [9, 10].
2. Proliferative stage—This stage takes place from a few days to the first couple of weeks following the injury. Blood vessels develop, and fibroblasts proliferate to produce matrix that fills the gap in the tendon.
3. Remodeling stage—This stage is characterized by decreasing cellularity and realigning of the matrix for an enhanced responds to the forces applied to the tissues and happens weeks to months after the initial injury [9, 10].

Unlike healing in skin and other organs where blood supply is usually ample for substrate delivery and cellular localization, tendons are avascular with discrete vascular leashes called vinculae occuring sparsely along their length.

6.3 Treatment for Tendon Injuries

The most common treatment strategy for complete tendon laceration is surgery, in which surgeons suture the lacerated tendon back together. Typically, a standardized technique with multiple strands of suture running through the core of the divided tendon and a running epitendinous repair is carried out. Early active motion is important to recovery of tendon function. After surgical repair, patients are given routine postsurgical tendon care and rehabilitation, which may include immobilization for the initial 2–6 weeks except for occupational therapy-guided joint manipulation to prevent stiffness (for example a modified Duran protocol). After this time period, the patient would be encouraged to start active motion. The balance between immobilization and occupational therapy is a challenge as too early active range of motion (ROM) may cause re-rupture, while delayed active ROM promotes significant risks of adhesion formation [11]. A fair number (5–20%) of patients require a second surgery due to re-rupture, and many patients develop adhesions, which interfere with the gliding motion of tendons that is required for joint motion. Tendon adhesions compromise function and limit the mobility of the joint. The traditional method of surgically treating tendon injuries is effective (tenolysis) but has significant risks of rupture and adhesion formation; thus it is not optimal [11].

6.4 Gene Therapy for Tendon Injuries

Damaged tendons heal poorly; as outlined above their surgical repair often ends in unpredictable rupture or impaired extremity motion due to stiffness or adhesions. Moreover, tendons, particularly those covered by an intrasynovial sheath, have very limited vascular supply, lack cellularity, and have low growth factor activity [11]. Several investigators have studied gene therapy as a potential solution to the poor tendon healing environment [11]. There are mainly three models that have been used: the Achilles tendon [12], the digital flexor tendon [13–16], and the patellar tendon of the knee [12, 17]. Among all models, various delivery methods, vectors, and manipulation of the healing environment have been performed, and all studies have been conducted in animals, mainly mice, rats, chicken, and dogs.

Initial studies focused on measuring the level of expression of the different genes, while later studies focused on manipulation of the healing environment to assess the improvement of healing strength and reduction of adhesion formation [17]. The delivery of growth factors, such as vascular endothelial growth factor (VEGF), basic fibroblast growth factor (bFGF), and platelet-derived growth factor-B (PDGF), to the tendons has demonstrated enhanced healing strength and reduced adhesion formation and rupture rate in animal studies.

Lou et al., evaluated rat patellar tendon healing using immunohistologies following implantation of PDGF-B gene through a vector. By 8 weeks, PDGF-B was expressed only at the wound site in both groups. Indirect fluorescence immunohistochemistry was used to evaluate angiogenesis, and a significant increase in the vascularity of the PDGF-B gene-treated patellar tendon wounds when compared to control tissues was seen 1 week postsurgery; however, there were no significant differences between the two groups at 4 and 8 weeks after surgery [13].

Nakamura et al., demonstrated early angiogenesis in its first week with the introduction of PDGF-B cDNA using HVJ liposomes and rat patellar tendon healing. This was associated with increased collagen deposition at 4 weeks and increased matrix deposition 1 and 4 weeks postsurgery. However, by 8 weeks, no differences between the PDGF-B group and control group with respect to angiogenesis, collagen deposition, and matrix staining were seen [18, 19].

In 2004, two of us (XTW and PYL) and Tang published a study on genetic modification of tenocytes by transferring the PDGF-B gene using plasmid/lipofectamine to tenocytes and investigated its effect on the expression of the PDGF gene and the type I collagen gene in an in vitro tenocyte culture model [17, 20]. By transferring the PDFG-B gene, we found that the levels of expression of the type I collagen gene by tenocytes were increased significantly [17, 20]. Next, we investigated the role of adeno-associated serotype virus 2 (AAV2) as a potential viral vector for gene transfer. In particular, we transferred the exogenous bFGF gene to proliferating tenocytes by AAV2 vectors and investigated its effects on the expression of the collagen genes in an in vitro tenocyte model. We found that the level of expression of the bFGF gene was increased significantly after gene transfer and subsequently, compared to controls [21]. Later we sought to identify the most efficient AAV vector for gene transfer. We tested the different serotypes of AAV in chicken primary tenocytes. We found that gene transfer to tenocytes by AAV2 is more efficient than any other serotypes [22, 23].

After the success of the above in vitro studies, we pursued the application of this model in vivo. Our collaborating investigators, Tang and colleagues, in Nantong University, China, transferred bFGF gene through the AAV2 vector to injured digital flexor tendons in chickens and investigated its effects on the healing strength of the tendon and on adhesion formation. This study demonstrated that the transfer of bFGF gene transfer significantly increased the healing strength during the critical tendon-healing period and did *not* increase adhesion formation [7]. Moreover, there were no apparent local or systemic toxicities in the chickens 12 weeks posttreatment.

To further investigate the role of growth factors in vivo, we compared AAV2-bFGF to AAV2-VEGF and placebo in chickens. In the two experimental groups, AAV2-bFGF or AAV2-VEGF was injected proximally and distally into four sites of the tendons immediately after tendon transection. Four and six weeks later, the tendons were harvested and were subjected to load-to-failure testing and work of digital flexion in a tensiometer. At 4 weeks, compared with the non-injection control tendons, the tendons injected with AAV2-bFGF and AAV2-VEGF had significantly greater tensile strength of 140% and 210% ($p < 0.01$, $p < 0.001$, respectively). At week six, the tendon strengths after injection of the AAV2-bFGF and the AAV2-VEGF were increased significantly compared with the control group. At week six, work of digital flexion (i.e., the energy required to flex the operated digits) was not significantly increased in the AAV2-bFGF or AAV2-VEGF injected tendons compared with the control tendons. We concluded that AAV2-bFGF and AAV2-VEGF could effectively improve the healing strength of the injured digital flexor tendon in an in vivo animal model. The expression of transgene started to increase at week 1–2 after treatment, but the transgene expression gradually decreased 6–8 weeks after treatment and could not be detected at week 12 and 16 [23].

6.5 Gene Therapy to Prevent Adhesion Formation

Adhesions are another major problem of tendon healing, which decreases tendon gliding and hampers active motion of the limbs. Two of us (YLZ and YFW) have aimed to use gene therapy to decrease adhesion formation around the tendon.

Nanoparticles are non-viral vectors and have high biocompatibility and apparent biosafety. The use of nanoparticles as gene delivery systems is ideal for treatment of damaged tendons, because this system has low toxicity and high biocompatibility. In 2012, we started to use this delivery system in tendon gene therapy and made preliminary reports of using this novel vector in the tendon in recent years [15, 16]. We delivered the TGF-β1 miRNA plasmid using this system to prevent flexor tendon adhesion formation. We choose poly(D,L-lactide-co-glycolide) (PLGA) to load nanoparticles using a double-emulsion water-oil-water solvent evaporation technique. PLGA has controlled release characteristics and great biological compatibility, which releases its contents over extended time periods. These features render it as an ideal tool for delivery of gene to the tendons to prevent adhesions, which typically developed weeks after surgical repair. We used a cationic polymer, polyetherimide (PEI), to modify nanoparticles to obtain PEI-modified PLGA nanoparticles. PEI is a cationic polymer, which can be used as a bridge to adsorb nucleic acids onto the surface of nanoparticles. Then PEI-modified nanoparticle solution was added to the plasmid solutions and vortexed gently to obtain PEI-modified PLGA nanoparticle/plasmid complexes. The diameter of the spherical PEI-modified PLGA nanoparticle/plasmid complexes is about 93 nm. Due to their nano-size range, polymeric nanoparticles containing genes of interest penetrate the cell membrane via endocytosis. PEI-modified PLGA nanoparticles were shown to escape from the endolysosomal compartment to the cytoplasmic compartment [16].

We recorded transfection efficiency of PEI-modified PLGA nanoparticles incorporating plasmids expressing enhanced green fluorescence protein (GFP) to the cultured tenocytes and in a chicken model. In cultured tenocytes from chicken, 45% of the cells expressed EGFP in the PEI-modified PLGA nanoparticle/plasmid complexes group 48 h after transfection. After injection of the nanoparticles into the flexor digitorum profundus tendon of the chicken, the nanoparticle/plasmid complex-treated tendons had significantly more EGFP-positive cells than the lipofectamine/pEGFP mixture-injected tendons ($p < 0.01$) at 1 week after surgery, and nanoparticles can ensure high-level transgene expression at least over a 4-week period. The transfection efficiency of PEI-modified PLGA nanoparticle/plasmid complexes in the injection region can reach about 39% at 1 week postsurgery and about 20% at 4 weeks postsurgery.

We also evaluated effectiveness of this nanoparticle system in limiting adhesions in chicken flexor tendons. Nanoparticles containing the TGF-β1 miRNA plasmids, aiming at inhibiting the TGF-β1 expression, were injected to transect digital flexor tendons of chickens. We harvested operated toes 6 weeks later; the breaking strength of the repaired tendons and the energy required to flex the toes were tested. We also evaluated the morphology of healing status and adhesions around the tendon. The gene and protein expression was evaluated by real-time PCR and immunoblotting. At the end of 6 weeks, the work of flexion of the operated toes after treatment with the PEI-modified nanoparticle/TGF-β1 miRNA plasmid was significantly less as compared with that treated with the PEI-modified nanoparticles/mock plasmid group and in non-injection control toes ($p < 0.05$ or $p < 0.01$). The strength of the repaired tendons treated with the PEI-modified nanoparticle/TGF-β1 miRNA plasmid was significantly lower than that of the tendons treated with the PEI-modified nanoparticles/mock plasmid ($p < 0.05$) and that of the non-injection control tendons ($p < 0.05$). At 6 weeks, the adhesions in the tendons treated with the PEI-modified nanoparticle/TGF-β1 miRNA plasmid group also have a downward trend compared to those treated with PEI-modified nanoparticles/mock plasmid and the non-injection control toes. We found that PEI-modified nanoparticle/TGF-β1 miRNA plasmid complexes substantially silenced the TGF-β1 gene and protein expression. After 6 weeks, there was significant downregulation in TGF-β1 and type III collagen gene expression in the tendon treated with PEI-modified nanoparticle/TGF-β1 miRNA plasmid complexes compared with that of the tendon treated with PEI-modified nanoparticle/mock plasmid and the non-injection control tendons ($p < 0.05$). However, expression of type I collagen in the tendons treated with the PEI-modified nanoparticle/TGF-β1 miRNA plasmid complexes showed a downward trend compared with the tendons treated with PEI-modified nanoparticle/mock plasmid complexes and non-injection control tendons [24–26].

Our results demonstrated that PEI-modified nanoparticle/TGF-β1 miRNA plasmid complexes are able to inhibit the expression of TGF-β1 in a sustained manner. We anticipate that PEI-modified nanoparticle/plasmid complexes injected into the tendon are cleared slowly as the polymer degrades gradually. These features make this gene delivery method a practical approach in preventing tendon adhesion. Similarly, we have used AAV2 vectors harboring engineered miRNA plasmid (TGF-β1 miRNA plasmid) to prevent adhesions. The preliminary results in chicken flexor tendon injury and repair model treated with AAV2-TGF-β1 miRNA plasmid reveal its effectiveness in decreasing adhesion formation, but this treatment also decreased the tendon healing strength [24–26].

Conclusions

In summary, gene transfer to tendons is achievable, and the healing environment can be modulated in salutary fashion. The tendon healing strength can be augmented, and adhesion formation around the tendons can be decreased through gene therapy approaches, at least in animal studies.

References

1. de Jong JP, Nguyen JT, Sonnema AJ, Nguyen EC, Amadio PC, Moran SL. The incidence of acute traumatic tendon injuries in the hand and wrist: a 10-year population-based study. Clin Orthop Surg. 2014;6(2):196–202.
2. Clayton RA, Court-Brown CM. The epidemiology of musculoskeletal tendinous and ligamentous injuries. Injury. 2008;39:1338–44.
3. de Putter CE, Selles RW, Polinder S, Panneman MJ, Hovius SE, van Beeck EF. Economic impact of hand and wrist injuries: health-care costs and productivity costs in a population-based study. J Bone Joint Surg Am. 2012;94(9):e56.
4. Jones BH, Bovee MW, Harris JM 3rd, Cowan DN. Intrinsic risk factors for exercise related injuries among male and female army trainees. Am J Sports Med. 1993;21(5):705–10.
5. Jones B, et al. Epidemiology of injuries associated with physical training among young men in the Army. Med Sci Sports Exerc. 1993;25:197–203.

6. Knapik J, Cuthie J, Canham-Chervak M, et al. Injury incidence, injury risk factors, and physical fitness of U.S. army basic trainees, Ft. Jackson, SC, 1997. Aberdeen Proving Ground, MD: US Army Center for Health Promotion and Preventive Medicine; 1998. Epidemiological consultation report 29-HE-7513-98.
7. Knapik J, Hauret K, Arnold S, et al. Injury and fitness outcomes during implementation of physical readiness training. Int J Sports Med. 2003;24:372–81.
8. Knapik J, Sharp M, Canham-Chervak M, Hauret K, Patton J, Jones B. Risk factors for training-related injuries among men and women in basic combat training. Med Sci Sports Exerc. 2001;33:946–54.
9. Hildebrand KA, Frank CB. Ligaments: structure, function and response to injury and repair. In: Dee R, Hurst LC, Gruber MA, Kottmeier SA, editors. Principles of orthopaedic practice. 2nd ed. New York: McGraw-Hill; 1997. p. 109–17.
10. Woo SL-Y, et al. Anatomy, biology, and biomechanics of tendon and ligament. In: Buckwalter JA, Einhorn TA, Simon SR, editors. Orthopaedic basic science: biology and biomechanics of the musculoskeletal system. 2nd ed; 2000. p. 581–616.
11. Hildebrand KA, Frank CB, Hart DA. Gene intervention in ligament and tendon: current status, challenges, future directions. Gene Ther. 2004;11(4):368–78.
12. Nakamura N, et al. A comparison of *in vivo* gene delivery methods for antisense therapy in ligament healing. Gene Ther. 1998;5:1455–61.
13. Lou J, et al. BMP-12 gene transfer augmentation of lacerated tendon repair. J Orthop Res. 2001;19:1199–202.
14. Lou J, Manske PR, Aoki M, Joyce ME. Adenovirus-mediated gene transfer into tendon and tendon sheath. J Orthop Res. 1996;14:513–7.
15. Lou J, et al. *In vivo* gene transfer and overexpression of focal adhesion kinase (pp125FAK) mediated by recombinant adenovirus-induced tendon adhesion formation and epitendon cell change. J Orthop Res. 1997;15:911–8.
16. Goomer RS, et al. Nonviral *in vivo* gene therapy for tissue engineering of articular cartilage and tendon repair. Clin Orthop. 2000;379S:S189–200.
17. Wang XT, Liu PY, Tang JB. Tendon healing in vitro: genetic modification of tenocytes with exogenous PDGF gene and promotion of collagen gene expression. J Hand Surg Am. 2004;29(5):884–90.
18. Nakamura N, et al. Early biological effect of *in vivo* gene transfer of platelet-derived growth factor (PDGF)-B into healing patellar ligament. Gene Ther. 1998;5:1165–70.
19. Nakamura N, et al. Transient introduction of a foreign gene into healing rat patellar ligament. J Clin Invest. 1996;97:226–31.
20. Wang XT, Liu PY, Tang JB. Tendon healing in vitro: modification of tenocytes with exogenous vascular endothelial growth factor gene increases expression of transforming growth factor beta but minimally affects expression of collagen genes. J Hand Surg Am. 2005;30(2):222–9.
21. Wang XT, Liu PY, Xin KQ, Tang JB. Tendon healing in vitro: bFGF gene transfer to tenocytes by adeno-associated viral vectors promotes expression of collagen genes. J Hand Surg Am. 2005;30(6):1255–61.
22. Zhu B, Cao Y, Xin KQ, Wang XT, Summerhayes IC, Liu PY, Tang JB. Tissue reactions of adenoviral, adeno-associated viral, and liposome-plasmid vectors in tendons and comparison with early-stage healing responses of injured flexor tendons. J Hand Surg Am. 2006;31(10):1652–60.
23. Wang XT, Liu PY, Tang JB, Mizukami H, Xin KQ, Ozawa K, Ushijima H. Tendon healing in vitro: adeno-associated virus-2 effectively transduces intrasynovial tenocytes with persistent expression of the transgene, but other serotypes do not. Plast Reconstr Surg. 2007;119(1):227–34.
24. Wu YF, Mao WF, Zhou YL, Wang XT, Liu PY, Tang JB. Adeno-associated virus-2-mediated TGF-β1 microRNA transfection inhibits adhesion formation after digital flexor tendon injury. Gene Ther. 2016;23(2):167–75.
25. Yang J, Liu A, Han Y, Li Q, Tian J, Zhou C. Osteodifferentiation of mesenchymal stem cells on chitosan/hydroxyapatite composite films. J Biomed Mater Res A. 2014;102(4):1202–9.
26. Tang JB, Zhou YL, Wu YF, Liu PY, Wang XT. Gene therapy strategies to improve strength and quality of flexor tendon healing. Expert Opin Biol Ther. 2016;16(3):291–301.

7 Gene Therapy and Free Flaps for Therapeutic Reconstructive Surgery

Christopher R. Davis, Peter A. Than, and Geoffrey C. Gurtner

7.1 Introduction

The goal of successful oncological plastic surgery is complete tumor excision with reconstruction to a premorbid level, combined with adjuvant oncological modalities as required. However, despite aggressive surgical margins and adjuvant therapy, recurrent disease remains a potential postoperative complication after reconstructive surgery. Reconstruction of soft tissue or bony defects may be performed using a variety of techniques. The reconstructive armamentarium has been revolutionized in the past 30 years with the use of microvascular free flaps, with the first successful free flap reported in 1973 [1, 2]. Free flaps, discrete autologous tissue composites capable of distant transfer using their own blood supply [3], are composed of a variety of different tissues including the skin, fascia, fat, muscle, or bone. Free flaps are transferred with an afferent artery and efferent vein with an intact microvascular network. Tissue transfer offers innumerable reconstructive possibilities for patients with ablative soft tissue or bony defects. However, free flaps currently offer a purely reconstructive role, but provide no therapeutic oncological function.

Gene therapy may permit manipulation of free flaps to secrete local chemotherapeutic agents, thus fulfilling dual reconstructive and oncological roles [4]. We have coined the term *therapeutic flaps* to describe such modified free flaps in oncological surgery. This chapter highlights the research advances in this field and the potential future clinical applications of therapeutic flaps.

C. R. Davis · P. A. Than · G. C. Gurtner (✉)
Department of Surgery, Division of Plastic Surgery, Stanford University School of Medicine, Stanford, CA, USA
e-mail: Christopher.Davis4@nhs.net; pthan@stanford.edu; ggurtner@stanford.edu

G. Giatsidis (ed.), *Gene Therapy in Reconstructive and Regenerative Surgery*, https://doi.org/10.1007/978-3-319-78957-6_7

7.2 Reconstructive Surgery: Current Clinical Practice

Tumor resection and reconstruction is a common plastic surgical procedure, with over 50,000 free flaps performed per year [5]. Tumor resections warranting microsurgical reconstruction include breast cancer, sarcoma, and head and neck cancer. Autologous tissue is used to reconstruct the secondary defects. Examples include fasciocutaneous flaps using abdominal tissue supplied by the deep inferior epigastric artery perforator for breast reconstruction [6, 7], fasciocutaneous flaps using anterolateral thigh tissue for head and neck reconstruction [8, 9], muscles flaps using latissimus dorsi for sarcoma or breast reconstruction [10, 11], or osteomyocutaneous flaps using the fibula and surrounding tissue for head and neck reconstruction [12]. Malignant tissue is surgically excised with appropriate margins, and autologous tissue from an appropriate donor site is raised, detached from its anatomical blood supply and anastomosed to recipient vessels at a distant recipient location as the flap is inset to fill the ablative defect. Reconstruction may be performed as part of the primary procedure (immediate reconstruction) or as a secondary procedure (delayed reconstruction).

Clinical progress in the field of reconstructive microsurgery has been made since the introduction of microvascular free tissue transfer in the 1970s [1, 2] and perforator mapping in the 1980s [13]. Advances in vascular imaging [14] and preoperative mapping of perforators using Doppler [15] or computerized tomography angiography [16, 17] have augmented clinical practice. Free-flap reconstruction is now safe, elective and predictable, and capable of improving patients' quality of life after ablative surgery [18, 19]. Free flaps play a purely reconstructive role for patients and possess the ability to restore form and/or function to a premorbid level. Evidence suggests that free flaps may be able to provide an additional therapeutic role for oncological patients [4]. By applying gene therapy autologous tissue, tissue is converted from a purely reconstructive role into a combined reconstructive and therapeutic role. We have created the term 'Therapeutic Flap' for future nomenclature of this therapeutic and reconstructive tissue.

7.3 Gene Therapy

Gene therapy manipulates host genomes to produce desirable proteins. It offers the theoretical promise of delivering specific genes to produce desired proteins capable of targeting specific cells. Initial excitement toward gene therapy has been tempered by some disappointment at the clinical trial stage [20–22], where the majority of clinical trials have ceased at phase II [23].

Gene therapy may provide gene delivery if a patient is congenitally devoid of a specific gene. The same principles may offer therapeutic advantages in cancer therapy. In the oncological setting, the genome of autologous free flaps for reconstructing ablative defects may be manipulated by gene therapy. Laboratory evidence has demonstrated local chemotherapeutic release and tumor subsidence after oncological reconstruction using modified flaps akin to therapeutic flaps [4].

For successful gene therapy, transgene uptake into the host nuclei requires three distinct stages: cell entry through the phospholipid bilayer of the eukaryotic cell membrane (transduction), passage across the nuclear membrane (translocation), and genetic incorporation into the host genome (transcription). Effective transgene transport and incorporation into host nuclei can be augmented through a vector. Vectors are genetic carrier molecules and may be viral or non-viral.

7.3.1 Viral Vectors

Viral vectors command the majority of the gene therapy evidence in plastic surgery [23], due to established incorporation within the host genome through high transduction efficiency [24]. Examples of viral vectors include:

- Adenovirus
- Adeno-associated virus (AAV)
- Retrovirus
- Herpes simplex virus (HSV)

Viral vectors utilize the innate adaptive processes of viruses in order to maintain their genome within a host. However, viral vectors are expensive and complex to design. Adenovirus is a double-stranded DNA virus capable of delivering large quantities in vivo and in vitro. They are immunogenic and may be targeted by host immunity. During initial adenoviral research, AAVs were fortuitously identified as a contaminant with adenoviruses [25]. AAVs are replication-deficient single-stranded DNA viruses that co-localize with viral material. They lack any association with disease and have oncoprotective properties [26–28]. AAVs are also advantageous compared to other vectors due to their increased stability, potential for purification, and longer half-life [29]. Retroviral vectors are two single strands of RNA, which reverse-transcribe into double-stranded cDNA in the host genome and cause less immune stimulation, though are technically challenging to grow and concentrate. HSVs are a neurotropic double-stranded DNA virus capable of the largest genetic transfer of viral vectors but clinically suboptimal due to high levels of preexisting antibodies to HSV.

7.3.2 Non-viral Vectors

Non-viral vectors transport the desired DNA through charge-based interactions, after binding or enveloping of genetic material has occurred [30]. Examples of vectors of non-viral origin include:

- Naked DNA/plasmids
- Liposomes
- Basic proteins
- Polymers

Non-viral vectors transport large quantities of DNA and are inexpensive and relatively inert to the host immune response [31]. However, they have a lower transduction efficiency [24], transient gene transfer [32], and lower specificity to desired cells.

Barriers to overcome before clinical application of gene therapy include:

- Inability to target the vector solely to the desired organ or area of interest
- Inability to achieve sufficient levels of the expressed transgene to exert a clinical effect
- Transient expression of the delivered gene using some vectors
- Toxicity resulting from the immune response to the viral vector
- Limit of safe viral particle administration

Currently, no method of gene delivery completely overcomes these limitations. Novel approaches of gene therapy delivery techniques using microvascular free tissue transfer have demonstrated targeted delivery of vectors resulting in therapeutic transgene expression in the flap and the production of a measurable biological outcome [5, 29].

7.4 Principles of Therapeutic Flaps

Therapeutic flaps offer numerous advantages to patients undergoing surgical resection after cancer ablation. The tissue used for reconstruction is autologous; therefore transferring tissue from one anatomical site to another circumvents the need for immunosuppressive drugs avoiding inherent side effects. Furthermore, unlike synthetic materials such as silicone implants, capsular formation and implant-related complications—such as rupture or change of implant position—do not occur, thus negating the need for the further surgery. Autologous tissue has an indefinite lifespan. By priming the free flap with therapeutic vectors prior to inset, the flaps' local genome becomes a "protein factory" for desired proteins [33]. During the ex vivo period when the tissue is detached, the free flap can be perfused with extremely high titers of viral vector that would otherwise be unsafe in systemic dissemination. Prior to reanastomosis, the free flap can be flushed free of unincorporated viral particles [34]. The dual role of the flap as a reconstructive unit of form and function, as well as a therapeutic source, enables traditional reconstructive surgery to become both reconstructive and therapeutic.

The anatomical position of the reconstructive flap as a local source of anticancer therapy is optimal. Flaps fill oncological ablative defects, where tumors have been surgically resected. Therapeutic flaps provide adjuvant therapy to the precise anatomical sites at greatest risk of local recurrence, providing:

- Cytotoxicity to residual microscopic disease
- Prophylaxis against future disease
- Tumor clearance in recurrent disease

Therapeutic flaps have the potential to be tumor-specific and patient-specific, allowing precise therapeutic intent tailored to individual patients. Penetrance of the flap genome may permit longstanding genetic incorporation, allowing for targeted messenger molecules to turn genes on or off in response to the clinical need for chemotherapeutic therapy. As an additional safety mechanism, in the rare case of an adverse reaction, the flap can be removed at any stage and standard reconstruction performed, thus maximizing the safety of therapeutic flaps.

Laboratory models have demonstrated the therapeutic action of free flaps akin to therapeutic flaps. Dempsey et al. have demonstrated the therapeutic properties of an interleukin-12 (IL-12) secreting flap in a rodent model of breast cancer [4]. By transfecting rodent cells with a breast adenocarcinoma cell line, tumor presence can be monitored directly by volumetric calculation of palpable tumor mass or indirectly by quantification of beta human chorionic gonadotropin gene (β-HCG) levels secreted by tumor cells in concentrations directly proportional to tumor size. After flap harvest, detached abdominal tissue—equivalent to tissue used for breast reconstruction—undergoes ex vivo transduction with adenoviral vector encoding IL-12. Rodent models secreting IL-12 from their re-anastomosed flap had significantly less tumor burden than the control groups, where lacZ represented an inactive placebo equivalent. Nine days postoperatively, tumor burden with IL-12 releasing flaps was significantly less than controls, as demonstrated by lower tumor volumes and reduced β-HCG concentration.

Systemic toxicity—a recognized complication of in vivo IL-12 administration [35–37]—did not occur in rodents with transfected flaps locally secreting IL-12 [4]. Histological examination of sections of the liver, spleen, and lung demonstrated normal structural architecture, while serum liver enzymes remained within the normal range.

The theoretical benefits from using therapeutic flaps as a method of gene delivery are:

- Highly efficient transduction of cells within the flap [4, 5].
- Protein production is targeted to the tumor bed—the precise location where positive margins, subsequent disease, and synchronous tumors occur.
- Minimal systemic transduction of uninvolved tissues or organs [4, 5].
- High titers of vectors may be safely administered because systemic delivery of the virus is limited [4, 5].
- Reversible procedure, since the flap can be removed if the patient experiences an adverse reaction to the viral vector or the transgene.

This form of biologic brachytherapy overcomes many of the challenges of gene therapy. Modified free flaps have shown promise in rodent models delivering local cancer immunotherapy to a tumor bed [4] and in delivering antimicrobial peptides (AMP) in models of chronic infection, demonstrating improved biological endpoints [33]. AMP delivery is an additional potential therapeutic benefit from genetically enhanced free flaps. This removes side effects from systemic therapy, such as drug toxicity [38], reduced concentration at the desired site, and patient compliance [39]. Recent attempts to overcome these shortcomings include refined antibiotic selection and combination therapy [39], incorporation of a drug within a liposome [40], and antibiotic-releasing beads [41]. Despite the initial promise of impregnated cement, two randomized controlled trials describe no clinical improvement in open fractures [42] and chronic osteomyelitis [43]. Drug elution from antibiotic-releasing beads is short term and provides high titers in the transient postoperative period only. Modified free flaps in reconstructive surgery, using the same principles as therapeutic flaps, could provide a more sustained release of targeted antimicrobial therapy from an established autologous source. This is supported by evidence of AMP release with bactericidal activity from keratinocytes within cultured skin after adenoviral cutaneous gene therapy [44].

7.5 Clinical Application of Therapeutic Flaps

The optimum clinical scenario for a patient undergoing contemporary free-flap surgery using a therapeutic flap, with ex vivo gene therapy prior to reconstruction, would result in the following:

- Genetic uptake at the flap and local protein expression.
- Chemotherapeutic protein release to the local environment.
- Longstanding transgene incorporation.
- Therapeutic gene(s) would respond to subsequent stimulation and turn on or off according to treatment needs of the patient and tumor.
- No tumor recurrence.

Clinical scenarios are illustrated below using head and neck cancer (HNC) and breast cancer as models of disease suited to the therapeutic benefits from therapeutic flaps.

7.5.1 Head and Neck Cancer

HNC is therapeutically challenging, with a locoregional recurrence rate of 50% despite combination therapy of surgery, radiotherapy, and chemotherapy [45]. Patients with recurrent disease have limited options for further treatment, resulting in a poor prognosis. Such patients are therefore potential candidates for cancer gene therapy via therapeutic flaps. Plastic surgeons and allied surgical specialities are commonly involved in resection and reconstruction of surgical defects of head and neck tumors. Mutations in tumor suppressor genes (TSG) p53, p16, p21, p27, and retinoblastoma as well as oncogenes cyclin D1 and epidermal growth factor and its receptor (EGF-R) have been identified in head and neck squamous cell carcinoma (HNSCC). The most intensively investigated TSG is p53, an inducer of cell cycle arrest and apoptosis in health. Although the predictive value of p53 mutations for tumor recurrence is controversial, in vitro and in vivo studies using recombinant

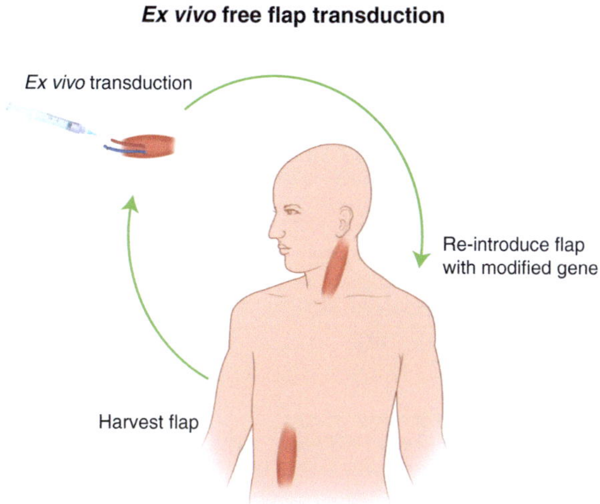

Fig. 7.1 Ex vivo transduction of free flaps in head and neck cancer reconstruction (From Ghali et al. [34])

adenovirus with wild-type p53 (Ad/p53) have induced apoptosis in various head and neck tumor cell lines and reduced the growth of established tumors [46]. A phase I clinical trial in patients with advanced locoregional recurrent HNC demonstrated the safety of Ad/p53 as an adjuvant therapy and also a median survival of 60% longer than that found in non-treated patients [47]. Phase II/III clinical trials are underway involving adjunctive Adp53 alongside radiotherapy, chemotherapy, and surgery in the treatment of HNC [48, 49].

With respect to the use of oncolytic viruses such as ONYX-015 in the treatment of HNSCC, phase I and II trials have demonstrated intra-tumor injection can be administered safely in patients with advanced HNC with moderate tumor response [50, 51]. A phase II trial that investigated combinations of ONYX-015 with cisplatin and 5-fluorouracil in 37 patients with head and neck cancer showed a partial response in 15 patients, all of which remained free of progression at 6 months [52]. These promising clinical trials add supportive evidence to the clinical promise of therapeutic flaps, where the reconstructive tissue is primed during the ex vivo phase, providing therapeutic benefit (Fig. 7.1).

Recruiting oncolytic viruses and ceasing malignant dysregulation of the cell cycle are an attractive means of targeting tumor cells via a therapeutic flap. However, it is imperative that the specificity of tumor targeting is solely to the cancer cells, and the integrity of the innate cells of patient and flap is preserved. Vascular supply with appropriate venous drainage is essential for maintenance of a viable reconstructive flap and therefore preservation of vascular endothelium. Methods to augment blood flow across microvascular anastomoses include topical and systemic administration of vascular endothelial growth factor [53, 54] and local injection of stem cells [55]. While tumor clearance and tissue reconstruction remain the primary goal, research attempts to augment flap survival must maximize patient safety without compromising oncological clearance.

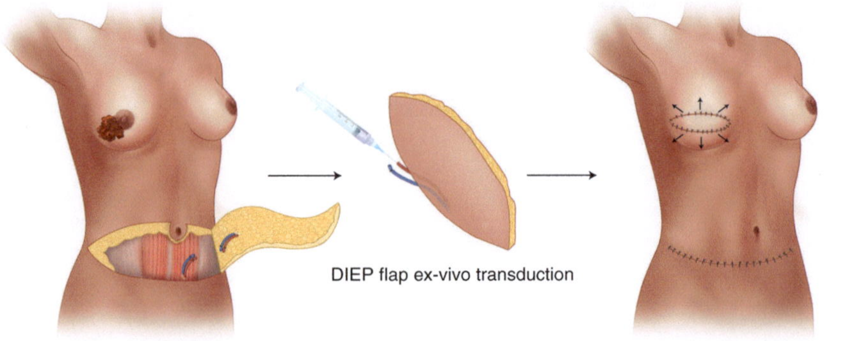

Fig. 7.2 Current surgical practice of raising a flap from the abdomen for immediate breast reconstruction following mastectomy (left image); proposed ex vivo transduction with vector (center image); local chemotherapeutic released after flap inset (right image) (From Dempsey MP et al. [4]). *DIEP* deep inferior epigastric perforator

7.5.2 Breast Cancer

Management of breast cancer is multimodal, with therapeutic contributions from combinations of surgery, chemotherapy, and radiotherapy. By considering the genetic basis of breast cancer, one may approach potential curative avenues through gene therapy. For example, gene therapy may provide therapeutic benefit in breast cancer by upregulation of tumor pro-apoptotic pathways [56]. Mutations in p53—a pro-apoptotic TSG—are associated with breast cancer pathogenesis [57, 58]. Mutated p53 prevents the apoptosis of cells with damaged DNA [59], thus promoting tumor propagation from renegade cells. Other genes involved in the apoptotic cascade with a potential role in breast cancer pathogenesis include bcl-2 [60, 61], Bax [62], and caspases [63]. Pro-apoptotic gene therapy may provide targeted tumor cell apoptosis, though currently there have been suboptimal tumor transduction [64]. Ablation of oncogene function is an additional pathway open to therapeutic genetic manipulation. Therapeutic flaps may therefore be inset after mastectomy to reconstruct a breast, capable of secreting anticancer factors that arrest dysregulated cell cycles in residual or recurrent tumor cells. By using a free flap as an autologous recipient for gene transfer, traditional breast reconstruction has the potential to become an oncologically therapeutic reservoir to reduce the risk of recurrent disease and target microscopic tumor cells by using therapeutic flaps (Fig. 7.2).

Conclusion

Surgical excision of tumors, combined with appropriate adjuvant or neoadjuvant therapy, may result in oncological recurrence. Autologous tissue, traditionally used for reconstruction only, is a potential therapeutic source to patients undergoing reconstructive surgery to ablative defects. Genetic manipulation of the autologous tissue during the ex vivo phase via gene therapy permits alteration of the free flap. The modified free flap releases therapeutically active proteins with the aim of eradicating potential tumor recurrences. This review has highlighted how reconstructive tissue can offer both reconstructive and therapeutic roles in the form of therapeutic flaps as a contemporary means of oncological therapy.

Contributors All authors contributed to the design, writing, and editing of this manuscript and granted approval of the final version.

Conflicts of Interest None of the authors have any conflicts of interest to declare.

References

1. Taylor GI, Daniel RK. The free flap: composite tissue transfer by vascular anastomosis. Aust N Z J Surg. 1971;43(1):1–3.
2. Daniel RK, Taylor GI. Distant transfer of an island flap by microvascular anastomoses. A clinical technique. Plast Reconstr Surg. 1973;52(2):111–7.
3. Mathes SJ. Muscle flaps and their blood supply. In: Aston SJ, Beasley RW, Thorne CW, editors. Plastic surgery. 5th ed. New York: Lippincott-Raven; 1997. p. 61–71.
4. Dempsey MP, Hamou C, Michaels J, Ghali S, Jazayeri L, Grogan RH, Gurtner GC. Using genetically modified microvascular free flaps to deliver local cancer immunotherapy with minimal systemic toxicity. Plast Reconstr Surg. 2008;121(5):1541–53.
5. Michaels J, Levine JP, Hazen A, Ceradini DJ, Galiano RD, Soltanian H, Gurtner GC. Biologic brachytherapy: ex vivo transduction of microvascular beds for efficient, targeted gene therapy. Plast Reconstr Surg. 2006;118(1):54–65.
6. Allen RJ, Treece P. Deep inferior epigastric perforator flap for breast reconstruction. Ann Plast Surg. 1994;32:32–8.
7. Blondeel PN, Boeckx WD. Refinements in free flap breast reconstruction: the free bilateral deep inferior epigastric perforator flap anastomosed to internal mammary artery. Br J Plast Surg. 1994;47:495–501.
8. Wei FC, Jain V, Celik N, et al. Have we found an ideal soft tissue flap? An experience with 672 anterolateral thigh flaps. Plast Reconstr Surg. 2002;109:2219.
9. Lin F, Grinsell D. Free anterolateral thigh perforator flap for head and neck cancer resection in a nonagenarian. Mod Plast Surg. 2012;2(1):1–4.
10. Tanzini. Sporo il nito nuova processo di aupertozione della menuelle. Riforma Med. 1906;22:757.
11. d'Este S. La technique de l'amputation de la mamelle pour carcinome mammaire. Rev Chir. 1912;45:164.
12. Cheng MH, Saint-Cyr M, Ali RS, Chang KP, Hao SP, Wei FC. Osteomyocutaneous peroneal artery-based combined flap for reconstruction of composite and en bloc mandibular defects. Head Neck. 2009;31(3):361–70.
13. Moon HK, Taylor GI. The vascular anatomy of rectus abdominis musculocutaneous flaps based on the deep superior epigastric system. Plast Reconstr Surg. 1988;82:815–32.
14. Saint-Cyr M, Schaverien M, Arbique G, et al. Three- and four-dimensional computed tomographic angiography and venography for the investigation of the vascular anatomy and perfusion of perforator flaps. Plast Reconstr Surg. 2008;121:772–80.
15. Blondeel PN, Beyens G, Verhaeghe R, et al. Doppler flowmetry in the planning of perforator flaps. Br J Plast Surg. 1998;51(3):202–9.

16. Rozen WM, Ribuffo D, Atzeni M, et al. Current state of the art in perforator flap imaging with computed tomographic angiography. Surg Radiol Anat. 2009;31:631–9.
17. Pacifico MD, See MS, Cavale N, et al. Preoperative planning for DIEP breast reconstruction: early experience of the use of computerised tomography angiography with VoNavix 3D software for perforator navigation. J Plast Reconstr Aesthet Surg. 2009;**62**(11):1464–9.
18. Dean C, Chetty U, Forrest AP. Effects of immediate breast reconstruction on psychosocial morbidity after mastectomy. Lancet. 1983;1:459–62.
19. Elder EE, Brandberg Y, Bjorklund T, et al. Quality of life and patient satisfaction in breast cancer patients after immediate breast reconstruction: a prospective study. Breast. 2005;14:201–8.
20. Vile RG, Russell SJ, Lemoine NR. Cancer gene therapy: hard lessons and new courses. Gene Ther. 2000;7:2–8.
21. Check E. Gene therapy: a tragic setback. Nature. 2002;420:116–8.
22. Check E. Cancer fears cast doubts on future of gene therapy. Nature. 2003;421:678.
23. Gene therapy clinical trials worldwide. Hoboken, NJ: John Wiley and Sons; 2007. www.wiley.com/legacy/wileychi/genmed/clinical
24. Kootstra NA, Verma IM. Gene therapy with viral vectors. Annu Rev Pharmacol Toxicol. 2003;43:413–39.
25. Blacklow NR, Hoggan MD, Sereno MS, et al. A seroepidemiologic study of adenovirus-associated virus infection in infants and children. Am J Epidemiol. 1971;94:359–66.
26. Bantel-Schaal U. Adeno-associated parvoviruses inhibit growth of cells derived from malignant human tumors. Int J Cancer. 1990;45:190–4.
27. Hermonat PL. Adeno-associated virus inhibits human papillomavirus type 16: a viral interaction implicated in cervical cancer. Cancer Res. 1994;54:2278–81.
28. Walz CM, Anisi TR, Schlehofer JR, et al. Detection of infectious adeno-associated virus particles in human cervical biopsies. Virology. 1998;247:97–105.
29. Michaels J, Dobryansky M, Galiano RD, et al. Ex vivo transduction of microvascular free flaps for localized peptide delivery. Ann Plast Surg. 2004;52:581–4.
30. Johnson-Saliba M, Jans DA. Gene therapy: optimising DNA delivery to the nucleus. Curr Drug Targets. 2001;2:371–99.
31. Wells DJ. Electroporation and ultrasound enhanced non-viral gene delivery in vitro and in vivo. Cell Biol Toxicol. 2010;26:21–8.
32. Wolff JA, Budker V. The mechanism of naked DNA uptake and expression. Adv Genet. 2005;54:3–20.
33. Ghali S, Bhatt KA, Dempsey MP, Jones DM, Singh S, Aarabi S, Butler PE, Gallo RL, Gurtner GC. Treating chronic wound infections with genetically modified free flaps. Plast Reconstr Surg. 2009;123(4):1157–68.
34. Ghali S, Dempsey MP, Jones DM, Grogan RH, Butler PE, Gurtner GC. Plastic surgical delivery systems for targeted gene therapy. Ann Plast Surg. 2008;60:323–32.
35. Brunetti-Pierri N, Palmer DJ, Beaudet AL, Carey KD, Finegold M, Ng P. Acute toxicity after high-dose systemic injection of helper-dependent adenoviral vectors into nonhuman primates. Hum Gene Ther. 2004;15:35.
36. Sung M, Chen SH, Thung SM, et al. Intratumoral delivery of adenovirus-mediated interleukin-12 gene in mice with metastatic cancer in the liver. Hum Gene Ther. 2002;13:731.
37. Portielje JE, Gratama JW, van Ojik HH, Stoter G, Kruit WH. IL-12: a promising adjuvant for cancer vaccination. Cancer Immunol Immunother. 2003;52:133.
38. Stiefeld SM, Graziani AL, MacGregor RR, Esterhai JL Jr. Toxicities of antimicrobial agents used to treat osteomyelitis. Orthop Clin North Am. 1991;22:439–65.
39. Holtom PD, Patzakis MJ. Newer methods of antimicrobial delivery for bone and joint infections. Instr Course Lect. 2003;52:745–9.
40. Hiemenz JW, Walsh TJ. Lipid formulations of amphotericin B: recent progress and future directions. Clin Infect Dis. 1996;22(S2):S133–44.
41. Henry SL, Galloway KP. Local antibacterial therapy for the management of orthopaedic infections: pharmacokinetic considerations. Clin Pharmacokinet. 1995;29:36–45.

42. Moehring HD, Gravel C, Chapman MW, Olsen SA. Comparison of antibiotic beads and intravenous antibiotics in open fractures. Clin Orthop. 2000;372:254–61.
43. Hedstrom SA, Lidgren L, Torholm C, Onnerfalt R. Antibiotic containing bone cement beads in the treatment of deep muscle and skeletal infections. Acta Orthop Scand. 1980;51:863–9.
44. Carretero M, Del Río M, García M, et al. A cutaneous gene therapy approach to treat infection through keratinocyte-targeted overexpression of antimicrobial peptides. FASEB J. 2004;18(15):1931–3.
45. Munro AJ. An overview of randomised controlled trials of adjuvant chemotherapy in head and neck cancer. Br J Cancer 1995;71:83–91.
46. Clayman GL, el-Naggar AK, et al. In vivo molecular therapy with p53 adenovirus for microscopic residual head and neck squamous carcinoma. Cancer Res. 1995;55(1):1–6.
47. Clayman GL, Frank DK, et al. Adenovirus-mediated wild-type p53 gene transfer as a surgical adjuvant in advanced head and neck cancers. Clin Cancer Res. 1999;5(7):1715–22.
48. Peng Z. Current status of gendicine in China: recombinant human Ad-p53 agent for treatment of cancers. Hum Gene Ther. 2005;16(9):1016–27.
49. Wilson JM. Gendicine: the first commercial gene therapy product. Hum Gene Ther. 2005;16(9):1014–5.
50. Ganly I, Kirn D, Eckhardt SG, et al. A phase I study of Onyx-015, an E1B attenuated adenovirus, administered intratumorally to patients with recurrent head and neck cancer. Clin Cancer Res. 2000;6(3):798–806.
51. Nemunaitis J, Khuri F, Ganly I, et al. Phase II trial of intratumoral administration of ONYX-015, a replication-selective adenovirus, in patients with refractory head and neck cancer. J Clin Oncol. 2001;19(2):289–98.
52. Khuri FR, Nemunaitis J, Ganly I, et al. A controlled trial of intratumoral ONYX-015, a selectively-replicating adenovirus, in combination with cisplatin and 5-fluorouracil in patients with recurrent head and neck cancer. Nat Med. 2000;6(8):879–85.
53. Li QF, Reis ED, Zhang WX, Silver L, Fallon JT, Weinberg H. Accelerated flap prefabrication with vascular endothelial growth factor. J Reconstr Microsurg. 2000;16:45.
54. Zhang F, Fischer K, Komorowska-Timek E, et al. Improvement of skin paddle survival by application of vascular endothelial growth factor in a rat TRAM flap model. Ann Plast Surg. 2001;46:314.
55. Lu F, Mizuno H, Uysal CA, Cai X, Ogawa R, Hyakusoku H. Improved viability *of* random pattern skin flaps through *the* use *of* adipose-derived stem cells. Plast Reconstr Surg. 2008;121(1):50–8.
56. Gomez-Navarro J, Arafat W, Xiang J. Gene therapy for carcinoma of the breast. Pro-apoptotic gene therapy. Breast Cacner Res. 2000;2:32–44.
57. Elledge RM, Allred DC. The p53 tumor suppressor gene in breast cancer. Breast Cancer Res Treat. 1994;32:39–47.
58. Berns EM, De Witte HH, Klijn JG, et al. Prognostic value of TP53 protein accumulation in human primary breast cancer: an analysis by luminometric immunoassay on 1491 tumor cytosols. Anticancer Res. 1997;17:3003–6.
59. White E. Tumor biology. p53, guardian of Rb. Nature. 1994;371:21–2.
60. Wu J. Apoptosis and angiogenesis: two promising tumor markers in breast cancer. Anticancer Res. 1996;16:2233–9.
61. Sierra A, Lloveras B, Castellsague X, et al. Bcl-2 expression is associated with lymph node metastasis in human ductal breast carcinoma. Int J Cancer. 1995;60:54–60.
62. Vakkala M, Lahteenmaki K, Raunio H, Paakko P, Soini Y. Apoptosis during breast carcinoma progression. Clin Cancer Res. 1999;5:319–24.
63. Thornberry NA, Lazebnik Y. Caspases: enemies within. Science. 1998;281:1312–6.
64. Takamiya Y, Short MP, Ezzeddine ZD, et al. Gene therapy of malignant brain tumors: a rat glioma line bearing the herpes simplex virus type 1-thymidine kinase gene and wild type retrovirus kills other tumor cells. J Neurosci Res. 1992;33:493–503.

Developing a Gene Therapy Clinical Trial: Tips and Tricks

8

Raman Mehrzad and Paul Y. Liu

Abbreviations

ERB Ethic Review Board
FDA Food and Drug Administration
IEC Independent Ethics Committee
IND Investigational New Drug
NIH National Institutes of Health
RAC Recombinant DNA Advisory Committee
REB Research Ethics Board

8.1 US Food and Drug Administration

The FDA is a federal agency of the US Department of Health and Human Services [1]. The FDA is responsible for protecting the public health by regulating and monitoring products, pharmaceutical drugs, supplements, food, biopharmaceutical, vaccines, etc. They supervise a range of areas to ensure that quality and safety are met for products that could have any health effects. They enforce other laws as well, notably Section 361 of the Public Health Service Act and associated regulations. The headquarters are located in Silver Spring, Maryland; however, they have many offices and several laboratories throughout the USA [1].

R. Mehrzad · P. Y. Liu (✉)
Department of Plastic Surgery, Rhode Island Hospital, The Warren Alpert Medical School of Brown University, Providence, RI, USA
e-mail: pliu@lifespan.org

G. Giatsidis (ed.), *Gene Therapy in Reconstructive and Regenerative Surgery*,
https://doi.org/10.1007/978-3-319-78957-6_8

8.1.1 Investigational New Drug

The federal law requires that a new drug, including gene therapy, should be the subject of an approved marketing application before being distributed and used across the USA. Since a sponsor will likely want to ship the investigational drug to clinical investigators in different states, it must seek an exemption from the legal requirements. Technically, that is what the IND is for, i.e., the means by which the sponsor obtains this exemption from the FDA [2].

The primary objective for the investigators is to show that early preclinical studies have determined that the product is reasonably safe for use in humans, as well as justifying that the drug exhibits pharmacological activities for commercial development. The applicant needs to show collected data and necessary information to confirm that the product will not expose humans to unreasonable risks when used in early-stage clinical studies [2].

There are two different IND categories, commercial and noncommercial, and there are three IND types [2]:

1. *Investigator IND*—This type of IND is submitted by physicians who both initiate and conduct an investigation and under whose immediate direction the investigational drug is ordered or dispensed.
2. *Emergency use IND*—This type of IND is submitted to allow the FDA to expedite the application process when time is of concern in emergency situations.
3. *Treatment IND*—This third type of IND is submitted for experimental drugs with promising clinical testing for serious or life-threatening conditions, while the final work is conducted and the review of the FDA takes place.

8.1.2 IND Application

The FDA has strict guidelines on what data needs to be included in the application. Typically, preclinical animal studies that have assessed pharmacological activity and acute toxicity potential are required [2]. In particular, there are three areas that need to be covered:

1. Clinical protocol and investigator information—A detailed protocol of the proposed study is required, portraying the trial design, dosages, risks, etc. Moreover, a risks assessment needs to be done to assess whether patients will be exposed to unnecessary hazard. This section should also include information and qualifications of the principal and other investigators to ensure they are qualified to fulfill their clinical trial duties. Moreover, information about informed consent and IRB needs to be included.
2. Manufacturing information—Information regarding manufacturer, stability, composition, and controls used for manufacturing the proposed substance and the drug product should be added here.

3. Toxicology studies and animal pharmacology—Information and preclinical data on all animal studies needs to be included in this session, especially regarding the toxicology, safety, and outcomes of the drug in animals to assess whether the product is safe for testing in humans. Information regarding previous use of the drug in humans should/could also be added here if applicable.

8.1.3 Specific Application Requirements

Although the above covers the three broad areas of the applications, below is a more detailed overview of what needs to be included. However, please note that each drug has their specific requirements and may need more or less detailed than what is outlined below. Therefore, we encourage all investigators to check the FDA website and guidelines before preparing for the IND hearing. Moreover, the forms and requirements detailed below may be subject to change [2]:

1. Cover sheet (FDA Form-1571)
2. Table of contents
3. Introductory statement and general investigational plan
   (a) Ordinarily, 2–3 pages should suffice. The information requested here is intended to place the developmental plan for the drug into perspective and to help FDA anticipate sponsor needs.
4. Investigator's Brochure
5. Protocols
   (a) Your detailed phase 1, 2, or 3 study protocol
6. Chemistry, manufacturing, and control information—information regarding, but not limited to:
   (a) Drug substance
   (b) Drug product
   (c) Human risks
   (d) Toxicology
   (e) Name of manufacturer
   (f) Preparation of the drug
   (g) Analytic methods to assure identity, quality, purity, and strength of the drug
7. Pharmacology and toxicology information
   (a) Pharmacology and drug distribution
   (b) Toxicology data
   (c) Toxicology—full data tabulation
   (d) Toxicology—GLP certification
   (e) Toxicokinetic data
   (f) Animal studies
   (g) Pharmacokinetic
   (h) Pharmacodynamics
   (i) Monitoring of effects
8. Precious human experience with the investigational drug
9. References

8.1.4 Pre-IND Meeting

Before submitting the IND application, the investigators could request a pre-IND meeting [3]. The purpose of a pre-IND meeting is for investigators to seek guidance from the FDA to optimize and improve their protocol before they submit the actual IND application. The investigator submits their protocol and suggested application for the FDA to review, and the agency will provide advice and give valuable feedback on the application, as well as answering any questions regarding the drug development. The FDA does not require these meetings; however, they are recommended because they can confirm the requirements of the development process [3]. It may be beneficial to meet with the FDA early in the drug development process so that any recommendations and questions could be answered and integrated into the development program [3].

8.1.5 Time Frame

After a sponsor has submitted an IND application, a wait of 30 calendar days has to be made before initiating any clinical trials. FDA will during this time review the IND for safety to assure that research subjects will not be subjected to unreasonable risk [2].

8.2 The Recombinant DNA Advisory Committee

The RAC is a federal advisory committee that reviews gene therapy trials and provides recommendations to the NIH director related to basic and clinical research involving recombinant or synthetic nucleic acid molecules. Since gene therapy research involves techniques that are relatively new with characteristics that may not be well-defined and have long-term risks and benefits that have not clearly been established, they may raise ethical, medical, and social considerations worthy of special attention and public discussion [4].

Studies that involve human gene transfer, i.e., the transfer into human research participants of either recombinant nucleic acid molecules, or DNA or RNA derived from recombinant nucleic acid molecules, or synthetic nucleic acid molecules, are reviewed by the RAC. The RAC oversees and examines these issues in a setting where public input and comment are encouraged [4].

The RAC committee consists of 21 national experts in various fields of medicine, science, ethics, genetics, and patient perspectives that review the considered gene therapy trial to consider the current state of knowledge and technology [4].

8.2.1 Who Needs to Submit Their Protocol to the RAC?

Research involving investigation of humans of material containing recombinant or synthetic nucleic acid molecules developed with NIH funds must be reviewed by the RAC, even if there are completely private funded sources, but the trial will be conducted in an institution that receives any NIH money for research with recombinant or synthetic nucleic acid molecules. Moreover, you may still submit your protocol to the RAC even if the above criteria do not apply to your research. It will provide feedback on your trial, design, and protocol by experts in the field and help improve the safety and science of your trial [5].

You are exempted from the RAC review process if the clinical study involves the administration of recombinant or synthetic nucleic acid molecules in which "induction or enhancement of an immune response to a vector-encoded microbial immunogen is the major goal, such an immune response has been demonstrated in model systems, and the persistence of the vector-encoded immunogen is not expected" [5].

We encourage all investigators to carefully read through the website of RAC before they start their application.

8.2.2 Specific Application Requirements

Below are the general requirements for the RAC application [5]:

1. Cover letter—(1) The letter should state that the documentation submitted complies with the requirements stated in RAC's Appendix M-I-A, "Requirements for Protocol Submission"; (2) identify the IBC and Institutional Review Board (IRB) at the proposed clinical trial site(s) responsible for local review and approval of the protocol; and (3) acknowledge that no research participant will be enrolled (see definition of enrollment in Section I-E-7 of the *NIH Guidelines*) until the RAC review process has been completed (see Appendix M-I-B, "RAC Review Requirements") and IBC approval (from the clinical trial site), IRB approval, and all applicable regulatory authorizations have been obtained.
2. Scientific abstract
3. Nontechnical abstract
4. Proposed clinical protocol—including tables, figures, and relevant manuscripts
5. Proposed informed consent document
6. *Curriculum vitae* or biosketch of key study personnel

8.2.3 Submission Information

By US mail, send it to:

Attn: Biotechnology Program Advisor NIH Office of Biotechnology Activities National Institutes of Health, 6705 Rockledge Drive, Suite 750, Bethesda, Maryland 20892-7985

By overnight express, send it to:

Attn: Biotechnology Program Advisor NIH Office of Biotechnology Activities National Institutes of Health, 6705 Rockledge Drive, Suite 750 Bethesda, Maryland 20817-1814

8.2.4 After Submission

They will confirm the receipt of the protocol within 3 business days. If anything needs to be edited or is missing, then they will contact you immediately. You will be notified of the outcome of the initial review after 15 business days from submission of all the information required and also notified whether your protocol has been selected for public RAC review [5].

8.2.5 Public RAC Review

In case your protocol has been selected for public review, RAC members will be assigned to do an in-depth written review of the protocol and will pose a series of questions about the trial. The person who is responsible for the trial will be asked to make a 15–20 min presentation about the protocol at the RAC meeting [5].

8.2.6 After the RAC Review

Once you start enrolling patients and start the trial, you need to, within 20 business days, submit the following information:

1. A copy of the informed consent document approved by the IRB
2. A copy of the protocol approved by the IBC and IRB
3. A copy of the final IBC approval from the clinical trial site
4. A copy of the final IRB approval
5. A brief written report that includes the following information: (1) how the investigator(s) responded to each of the RAC's recommendations on the protocol (if applicable) and (2) any modifications to the protocol as required by FDA
6. Applicable NIH grant number(s)
7. The FDA IND number
8. The date of the initiation of the trial

In addition, investigators must submit an annual report, providing specific information about the trials [5].

8.2.7 RAC Timing

The RAC has four meetings per year and their website clearly states the dates and times. The review of the RAC application will be conducted within these meetings, and to be able to make it to a specific meeting, you need to submit your application 8 weeks prior to one of those meetings. The specific deadlines are stated on their website [4].

8.2.8 Institutional Review Board (IRB)

The IRB, also known as the Independent Ethics Committee (IEC) or Ethic Review Board (ERB) or Research Ethics Board (REB), is a committee in medical research institutions in the USA that has been designed to approve, review, and monitor scientific research, especially ones that involve humans. Here, a committee is assigned to review each trial, perform a risk-benefit analysis to the subjects, and assess whether the trial should be performed or not. In the USA, all gene therapy trials involving humans need to be approved by an IRB of that institution (or in some cases use of a commercial IRB may be possible) before the trial is began. Different institutions have different IRB protocol requirements, and investigators are encouraged to review the specifics for their own institution [6].

Other countries may or may not have similar committees. Nevertheless, investigators are encouraged to make sure that their gene therapy trial is meeting the rules and regulations of their institution before they conduct the research.

8.2.9 Strategies

Conducting a gene therapy trial requires careful planning and structure. However, the above information should be helpful to understand the steps needed to be taken before conducting the trial.

The first step is to design your protocol and study trial. Be specific on what you are going to do and how. Gather all of your preclinical data, and make sure you have safety and toxicology data on your preclinical animal studies. Once this part is done, we suggest you start working on the RAC application. We suggest starting with the RAC and then going forward with the IND because the RAC typically takes less time to get approved and some of the information on the RAC application can usually overlap between and you avoid duplicating your work on some parts. Moreover, by working with the RAC application, you may come up with question that you would like to ask at the pre-IND meeting (see below).

Once your RAC application is submitted, you should start working on the IND application. Remember that some information may be the same as in the RAC application, so you can use that information to avoid unnecessary work. During the initial work of the IND application, we suggest requesting a pre-IND meeting. In this meeting you can seek guidance from the FDA to optimize and improve their protocol before you submit the actual IND application and potentially avoid delays in the process. Moreover, this is an excellent chance for you to ask necessary questions and get valuable feedback on your protocol and study design. Again, this is not required by the FDA, but highly recommended.

The IRB will typically not approve your application until the IND and/or RAC is approved; however, this may vary from one institution to another, and we suggest each investigator to look into their own institutions' specific requirements.

8.3 In Summary

1. Develop your study protocol.
2. Gather all preclinical data.
3. Start with the RAC application.
4. Submit the RAC application.
5. Start with the IND application.
6. Request a pre-IND meeting.
7. Continue/edit your IND application according to the pre-IND meeting.
8. Submit your IND application.
9. Start with the IRB protocol.
10. Once IND application is approved, submit the IRB protocol.

References

1. FDA. Available at: www.fda.gov.
2. Investigational new drug application. Available at: http://www.fda.gov/drugs/developmentapprovalprocess/howdrugsaredevelopedandapproved/approvalapplications/investigationalnewdrugindapplication/default.htm.
3. Pre-IND meeting. Available at: http://www.fda.gov/downloads/Drugs/.../Guidances/ucm153222.pdf.
4. National Institution of Health, Office of Science Policy: RAC. Available at: http://osp.od.nih.gov/office-biotechnology-activities/biomedical-technology-assessment/hgt/rac.
5. RAC application – Q/A. Available at: http://osp.od.nih.gov/sites/default/files/NIH_Review_Process_HGT.pdf.
6. The Office of Human Research Protection. Institutional review board guidebook. "Chapter 3, Section A: Risk/benefit analysis." pp. 1–10.

Future Directions in Reconstructive and Regenerative Surgery

9

Giorgio Giatsidis

9.1 Innovations Transforming the Future of Gene Therapies

The CRISPR/Cas9 system for genome editing and reprogramming, first described in 2015, has represented a substantial breakthrough in gene therapies [1, 2]. The system adopts a prokaryotic defense mechanism to edit target genes in cells of interest with significantly higher efficiency and efficacy (thanks to its single guide RNA, gRNA) compared to previous methods that used zinc-finger nucleases and transcription activator-like effector nucleases [2, 3]. Introduction of CRISPR/Cas9 components into target cells can use either viral or non-viral systems, both of which have been described in previous chapters. Another advantage of the CRISPR/Cas9 system is that, by using multiple gRNAs, it is possibly to simultaneously modify and modulate (activation or inhibition) several target genes within the same cell. This capacity represents an important step forward for gene therapies. It is also of primary interest in reconstructive and regenerative surgery as multiple factors need to be regulated in an orchestrated way to promote tissue repair and regeneration (e.g., during wound healing, where "monotherapies" have shown low efficacy). These improvements directly affect costs of gene therapies and research time (it has been estimated that CRISPR/Cas9 has already allowed a 40–80% improvement in costs/time and that these numbers will grow exponentially). Indirectly, this innovation has also affected the ability to screen/test a larger number of therapeutic approaches (e.g., combinations of factors, time, and space regulation of expression) to better optimize targeted strategies before their translation into patient care. At the beginning of 2018, several clinical trials using the CRISPR/Cas9 in patients had

G. Giatsidis
Division of Plastic Surgery, Department of Surgery, Brigham and Women's Hospital, Harvard Medical School, Boston, MA, USA
e-mail: ggiatsidis@bwh.harvard.edu

G. Giatsidis (ed.), *Gene Therapy in Reconstructive and Regenerative Surgery*,
https://doi.org/10.1007/978-3-319-78957-6_9

already been registered or started worldwide. Besides the classic direct therapeutic approach to treat established diseases or tissue injuries, the CRISPR/Cas9 system has been recently proposed as an indirect therapeutic method to edit the immunological system and create "universal" xenografts from genetically modified animals, that won't reject once transplanted in humans, or to induce immune tolerance in vascularized composite allotransplants (VCAs) [4–6]. The introduction of the CRISPR/Cas9 system has not changed but only expanded the clinical indications, the target tissues (e.g., skin, cartilage, bone, skeletal muscles, nerves, flaps), and the molecular/cellular targets (e.g., growth factors, transcriptional factors, etc.) of gene therapies in the field of reconstructive and regenerative surgery [7–9]. Ethical concerns are still a matter of debate and might strengthen in parallel with the increased opportunities provided by this system, despite its safer outcomes. Yet, this innovation helps overcome many technical, regulatory, and manufacturing challenges previously experienced by classic gene therapy approaches and is leading to more efficient and more cost-effective therapeutic solutions.

Specific micro-RNAs (miRNAs) can also be used to correct a pathologic condition [10–14]. miRNAs are short endogenous noncoding RNAs mediating post-transcriptional regulation of gene expression (induction of mRNAs degradation and inhibition of mRNAs translation), and modulate virtually all reparative-regenerative processes occurring in human tissues. Several studies have shown that a dysregulation of miRNAs is present in many pathologic conditions, such as impaired wound healing. Antisense microRNA inhibitors (antimiRs or miRNA mimics) can also therapeutically inhibit specific miRNA in target cells. miRNA can be delivered systemically or locally, using the same techniques previously discussed for gene therapies (viral and non-viral delivery systems), or similar methods. Yet, effective, stable, and targeted delivery still represents a major challenge of these experimental therapies. An advantage of miRNA-based gene therapies is that a single miRNA usually regulates and controls the expression of several target genes, often functionally related (gene networks) [11]. Hence, through miRNAs, it is possible to achieve a wider modulation of physiological and pathologic processes in comparison to what was previously possible through standard gene therapies. The effectiveness of miRNA-based gene therapies has already been proven in several preclinical studies on wound healing and tissue regeneration.

Conclusions

As of today, over 1500 clinical trials involving gene therapies have been performed worldwide. Pushed by biotechnological advances, the field of gene therapies is becoming increasingly more efficient, effective, affordable, and safe. These improvements are creating new research and clinical opportunities, are promoting growing interest and enthusiasm by industry, and are increasing the accessibility of regulatory pathways. Despite most treatments are still experimental and are at the early stages of optimization, there is an emerging trend towards the development of approaches that could have a broader reach in clinical care. The unique conditions that relate to the application of gene therapies to the field of reconstructive and regenerative surgery (e.g., localized delivery, time-

and space-limited or modulated action, possibility to control/regulate several target genes simultaneously, combined use with tissue-engineered scaffolds or stem cells, etc.) are likely to enable within the next decade the translation of a substantial number of treatments to patients.

References

1. McNutt M. Breakthrough to genome editing. Science. 2015;350(6267):1445. https://doi.org/10.1126/science.aae0479.
2. Pineda M, Moghadam F, Ebrahimkhani MR, Kiani S. Engineered CRISPR systems for next generation gene therapies. ACS Synth Biol. 2017;6(9):1614–26. https://doi.org/10.1021/acssynbio.7b00011.
3. Wang HX, Li M, Lee CM, et al. CRISPR/Cas9-based genome editing for disease modeling and therapy: challenges and opportunities for nonviral delivery. Chem Rev. 2017;117(15):9874–906. https://doi.org/10.1021/acs.chemrev.6b00799.
4. Puga Yung GL, Rieben R, Bühler L, Schuurman H-J, Seebach J. Xenotransplantation: where do we stand in 2016? Swiss Med Wkly. 2017;147(506):w14403. https://doi.org/10.4414/smw.2017.14403.
5. Feng W, Dai Y, Mou L, Cooper D, Shi D, Cai Z. The potential of the combination of CRISPR/Cas9 and pluripotent stem cells to provide human organs from chimaeric pigs. Int J Mol Sci. 2015;16(3):6545–56. https://doi.org/10.3390/ijms16036545.
6. Butler JR, Wang Z-Y, Martens GR, et al. Modified glycan models of pig-to-human xenotransplantation do not enhance the human-anti-pig T cell response. Transpl Immunol. 2016;35:47–51. https://doi.org/10.1016/j.trim.2016.02.001.
7. Duncan KM, Mukherjee K, Cornell RA, Liao EC. Zebrafish models of orofacial clefts. Dev Dyn. 2017;246(11):897–914. https://doi.org/10.1002/dvdy.24566.
8. Hainzl S, Peking P, Kocher T, et al. COL7A1 editing via CRISPR/Cas9 in recessive dystrophic epidermolysis bullosa. Mol Ther. 2017;25(11):2573–84. https://doi.org/10.1016/j.ymthe.2017.07.005.
9. Mofazzal Jahromi MA, Sahandi Zangabad P, Moosavi Basri SM, et al. Nanomedicine and advanced technologies for burns: preventing infection and facilitating wound healing. Adv Drug Deliv Rev. 2018;123:33–64. https://doi.org/10.1016/j.addr.2017.08.001.
10. Meng Z, Zhou D, Gao Y, Zeng M, Wang W. miRNA delivery for skin wound healing. Adv Drug Deliv Rev. 2017. https://doi.org/10.1016/j.addr.2017.12.011.
11. Santulli G. microRNA: medical evidence, Advances in experimental medicine and biology, vol. 888. Cham: Springer; 2015. https://doi.org/10.1007/978-3-319-22671-2.
12. Wu YF, Mao WF, Zhou YL, Wang XT, Liu PY, Tang JB. Adeno-associated virus-2-mediated TGF-β1 microRNA transfection inhibits adhesion formation after digital flexor tendon injury. Gene Ther. 2016;23(2):167–75. https://doi.org/10.1038/gt.2015.97.
13. Icli B, Nabzdyk CS, Lujan-Hernandez J, et al. Regulation of impaired angiogenesis in diabetic dermal wound healing by microRNA-26a. J Mol Cell Cardiol. 2016;91:151–9. https://doi.org/10.1016/j.yjmcc.2016.01.007.
14. Chang K-P, Lee H-C, Huang S-H, Lee S-S, Lai C-S, Lin S-D. MicroRNA signatures in ischemia-reperfusion injury. Ann Plast Surg. 2012;69(6):668–71. https://doi.org/10.1097/SAP.0b013e3182742c45.

MIX
Papier aus verantwortungsvollen Quellen
Paper from responsible sources
FSC® C105338

If you have any concerns about our products,
you can contact us on
ProductSafety@springernature.com

In case Publisher is established outside the EU,
the EU authorized representative is:
Springer Nature Customer Service Center GmbH
Europaplatz 3, 69115 Heidelberg, Germany

Printed by Libri Plureos GmbH
in Hamburg, Germany